"Like many compelling images, it may turn out to be a mirage, but for the first time in the history of science we can form a conception of what a complete scientific theory of the world will look like."

A new theory of the universe is emerging from recent discoveries in fundamental physics and cosmology. Some scientists now believe that all of nature is ultimately under the control of a single superforce, which weaves together space, time and matter into a unified field theory. Bringing together the frontiers of high energy particle physics and cosmology, the superforce explains what caused the big bang that gave birth to the universe, and how the foundations for the cosmic structures we now see were forged in the microseconds that followed. It also suggests a bizarre new idea – that space and time are in reality eleven-dimensional, with the unseen dimensions of space masquerading as nuclear and electromagnetic forces.

Writing with clarity and elegance Paul Davies, recently described in the *New York Times* as "one of the most adept science writers on either side of the Atlantic", relates the story of these awesome new advances: "With the superforce unleashed, we could change the structure of space and time, tie our own knots in nothingness and build matter to order. Controlling the superforce would enable us to construct and transmute particles at will, thus generating exotic forms of matter. We might even be able to manipulate the dimensionality of space itself, creating bizarre artificial worlds with unimaginable properties."

Paul Davies, himself a research scientist in this subject area, believes that we might soon realize man's age-old dream – a unified theory of all existence.

Paul Davies is Professor of Theoretical Physics at the University of Newcastle upon Tyne. He holds a doctorate from the University of London where he spent eight years as a lecturer in mathematics. His research has ranged across much of fundamental physics and cosmology, and he has acquired an international reputation as a science popularizer – **Other Worlds, The Runaway Universe, The Edge of Infinity** and **God and the New Physics** received world-wide acclaim. He also writes regularly for *The Guardian, The Economist* and a number of science periodicals, and frequently contributes to television and radio broadcasts.

By the same author

The Runaway Universe
Other Worlds
The Edge of Infinity
God and the New Physics

Student texts

Space and time in the modern universe
The forces of nature
The search for gravity waves
The accidental universe
Quantum mechanics

Technical

The physics of time asymmetry
Quantum fields in curved space (with N. D. Birrell)

The Search for a Grand Unified Theory of Nature

Paul Davies

UNWIN PAPERBACKS
London Sydney

First published in Great Britain by William Heinemann Ltd 1984
First published in Unwin Paperbacks 1985
Reprinted 1987

UNWIN HYMAN LIMITED
Denmark House
37–39 Queen Elizabeth Street
LONDON SW1 2QB
and
40 Museum Street, LONDON WC1A 1LU

Allen & Unwin Australia Pty Ltd
8 Napier Street, North Sydney, NSW 2060 Australia

Unwin Paperbacks with Port Nicholson Press
60 Cambridge Terrace, Wellington, New Zealand

ISBN 0–04–539006–1

Printed in Great Britain by
Cox & Wyman Ltd, Reading

ACKNOWLEDGEMENTS

The author acknowledges permission to quote as follows:
page 42 from *Physics and Philosophy* by Werner Heisenberg, published by Harper & Row, Publishers Inc, New York, 1958;
page 165 from W. K. Clifford's 1879 *Lectures and Essays* edited by L. Stephen and F. Pollock, published by Macmillan London and Basingstoke;
page 54 from *Dismantling the Universe* by Richard Morris, published by Simon & Schuster, New York 1984;
page 238 from *The Trouble with Thinking Backwards* by Ralph Estling, New Scientist 2 June 1983 pages 619-621.
page 238 from *Life Beyond Earth* by Gerald Feinberg and Robert Shapiro, published by William Morrow & Co Inc, New York.

ALSO AVAILABLE

ALAN TURING
The Enigma of Intelligence
Andrew Hodges
THE ARTIFICIAL FAMILY
R Snowden & G D Mitchell
THE CAUSES OF WARS
Michael Howard
THE CHANGE MASTERS
Corporate Entrepreneurs at Work
Rosabeth Moss Kanter
CHEATS AT WORK
An Anthropology of Workplace Crime
Gerald Mars
CHIMPANZEE POLITICS
Power and Sex among Apes
Frans de Waal
CONTEMPORARY FEMINIST THOUGHT
Hester Eisenstein
GOD'S BANKER
The Life and Death of Roberto Calvi
Rupert Cornwell
HAVING TO
The World of Oneparent Families
E Cashmore
A HISTORY OF WESTERN PHILOSOPHY
Bertrand Russell
THE INCREDIBLE EURODOLLAR
Or Why the World's Money System is collapsing (3rd Ed)
W P Hogan & Ivor F Pearce
JAPAN IN THE PASSING LANE
An Insider's Account of Life in a Japanese Auto Factory
Satoshi Kamata
THE LEFT HAND OF CREATION
The Origin and Evolution of the expanding Universe
John D Barrow & Joseph Silk
PHILOSOPHY IN THE TWENTIETH CENTURY
A J Ayer
THE PROTESTANT ETHIC AND THE SPIRIT OF CAPITALISM
Max Weber
RASTAMAN
The Rastafarian Movement in England
E Cashmore
SOCIALISM IN A COLD CLIMATE
John Griffith et al
STRATEGIES FOR WOMEN AT WORK
Janice LaRouche & Regina Ryan
THE SUBVERSIVE FAMILY
Ferdinand Mount
SUPERFORCE
The Search for a Grand Unified Theory of Nature
Paul Davies
100 BILLION SUNS
The Birth, Life and Death of the Stars
R Kippenhahn
IN THE UNDERWORLD
Laurie Taylor

Contents

Preface

The recent surge of public interest in fundamental physics is one of the more unusual social developments of our time. What is it about physics, with its impenetrable formulae and esoteric technical jargon, that attracts a wider audience? The answer, I believe, lies with its immense power to explain the world, together with the deep mystical element that characterizes much of 'the new physics'. Alone among the sciences, physics claims to be an all-encompassing discipline, its subject matter the whole universe. Through physics, all parts of the cosmos, from the elementary particles within atoms to the largest astronomical structures, can be incorporated into a single conceptual framework. The ability of physics to unify the strange and bewildering world about us cannot fail to be profoundly inspiring.

In this book I have described what could be the greatest triumph yet of the new physics – a complete theory of the universe, including its origin. This astonishing prospect stems from a series of major advances in our understanding of the fundamental forces that control all natural activity. Recent research is revealing the existence of a supreme superforce of which all other forces are but facets. Fresh discoveries have opened the way to a radical new concept of a unified universe born amid extreme violence, in which all physical structures are fashioned out of the primeval fire under the action of the superforce.

This revolution in our understanding of the cosmos is happening now. Indeed, as I write these words, reports are coming in about new advances and discoveries concerning some of the topics dealt with in the forthcoming chapters. I have written this book because I wish to share my excitement and enthusiasm about these awesome developments with the wider public. Although many of the concepts are difficult and abstract, I have endeavoured to expound them in informal, non-technical language as far as possible.

The professional scientist who writes a book of this type for a general readership faces a special sort of problem. He has a responsibility both to his profession to maintain accuracy and balance, and to the reader not to smother the essential excitement of the subject with endless caveats and cautions. The problem is even more acute when dealing with a rapidly developing subject at the forefront of scientific endeavour, for ideas and fashions can change almost overnight, and theories which may hold credence at the time of writing may well fall from favour even by the time the book emerges from the press. On top of this, opinions will often differ among the specialists themselves as to the degree of credence which should be attached to any particular current theory.

In this book I present what is largely a personal view of the subject. Many of the ideas are part of established physical science, others remain controversial, and some are highly speculative. No doubt a lot of my professional colleagues would strongly disagree with the weight that I may have attached to some of the more speculative ideas. The reader who is concerned about separating fact from conjecture should therefore exercise care. For example, the big bang theory of the origin of the universe is more or less accepted by all scientists. Many astronomers and physicists also accept the detailed theory about the nuclear processes that must have occurred during the first few minutes. However, there is now considerable interest in the much more exotic processes that occurred during the first one second. This is much shakier ground, with little or no observational evidence (as yet) to confirm the various theories. All discussion of this very early epoch must therefore be considered as conjectural.

Many other new and speculative theories are discussed in this book. One of these is the idea that spacetime has eleven dimensions. At the time of writing this theory is very popular with some theorists, and has a number of compelling features to commend it, but there is no experimental verification, and nobody accepts eleven-dimensional spacetime as established fact. Another important recent development is known as supersymmetry. Here the theory has reached a more advanced state of development, and many theorists speak strongly in favour of it. Nevertheless, supersymmetry remains a conjecture. The same may be said for the so-called grand unified theories, or GUTs, though here we could soon find some sort of experimental support.

The fact that ideas are speculative does not, of course, diminish our interest in them. Science thrives on speculation and conjecture. In some areas, such as cosmology, where experimental tests are difficult or even impossible, scientific controversy tends to involve confrontation between rival theories, rather than between theory and observation. Nevertheless, progress in understanding can still be made, by keeping to the rules of logic, and insisting on consistency with known physics.

What follows, then, is a model of the universe. It is not my invention, but one that is currently enjoying favour among some theoretical physicists and cosmologists, at least in many of its aspects. I personally believe that, although the model will undoubtedly change in the future, most of the essential ideas are correct, and will become more widely accepted in years to come.

While writing this book I have continued to enjoy many fruitful discussions with friends and colleagues. Much of the content reflects the beautiful insights that they have provided. I am especially indebted to my immediate colleagues at the University of Newcastle, Dr Stephen Bedding and Dr Ian Moss, who have greatly assisted me with technical details. I have also received important help and information from Dr Alan Guth and Professor Martin Rees.

P. D.

Technical note: Powers of ten notation

In this book we shall often meet very large and very small numbers. Rather than writing them out in full, the more convenient 'powers of ten' notation will be used, in which the number 10^n stands for one followed by n zeros. Thus:

one million	$= 10^6$	= 1 000 000
one billion	$= 10^9$	= 1 000 000 000
one trillion	$= 10^{12}$	= 1 000 000 000 000, etc.

Similarly, 10^{-n} stands for one divided by 10^n. Thus:

one millionth	$= 10^{-6}$	= 1/1 000 000
one billionth	$= 10^{-9}$	= 1/1 000 000 000
one trillionth	$= 10^{-12}$	= 1/1 000 000 000 000, etc.

I
The Unfolding Universe

Creation and the search for the superforce

Everybody likes adventure stories. One of the greatest adventures of all time is happening now, in the shadowy world of fundamental physics. The characters in the story are scientists and their quest is for a prize of unimaginable value – nothing less than the key to the universe.

The most important scientific discovery of our age is that the physical universe did not always exist. Science faces no greater challenge than to explain how the universe came to exist and why it is structured in the way it is. I believe that in the last few years that challenge has been met. For the first time in history we have a rational scientific theory of all existence. This revolutionary breakthrough represents an advance of unparalleled magnitude in our understanding of the world and will have profound repercussions for man's conception of the cosmos and his place within it.

These dramatic developments stem directly from a number of major advances made in fundamental physics over the last decade, especially in the area known as high-energy particle physics. On the experimental front, important discoveries are revealing for the first time deep relationships between subnuclear particles and the forces that lie buried within matter. But the advances in theoretical understanding are, if anything, even more spectacular. Two new conceptual schemes are currently forcing the pace. One goes under the name of 'grand unified theories', or GUTs. The other is called 'supersymmetry'. Together these investigations point towards a compelling idea, that all nature is ultimately controlled by the activities of a single *superforce.* The superforce would have the power to bring the universe into being and to furnish it with light, energy, matter, and structure. But the superforce would amount to more than just a creative agency. It would represent an amalgamation of matter, spacetime, and force into an integrated and harmonious framework that bestows upon the universe a hitherto unsuspected unity.

All science is essentially a search for unity. The scientist, by relating different phenomena in a common theory or description, unifies part of our bewilderingly complex world. What makes the recent discoveries so exciting is that, in theory, *all* natural phenomena can now be encompassed within a single descriptive scheme.

The search for a superforce can be traced to the early work of Einstein and others, who attempted to construct a *unified field theory*. A century before, Faraday and Maxwell had shown that electricity and magnetism are intimately related forces, which can be described by a unified electromagnetic field. The success of this description can be measured by the tremendous impact that radio and electronics – which derive from the electromagnetic field concept – have had on our society. The motivation has always been compelling to extend the unification process, and merge the electromagnetic field with other force fields, such as gravity. Who knows what extraordinary results that might bring?

The next step, however, did not prove so easy. Einstein's quest for a unified theory of the electromagnetic and gravitational fields was in vain, and it was not until the late 1960s that further progress on the road to unity was taken, when it was shown that electromagnetism can be mathematically combined with one of the nuclear forces (known to physicists as the weak force). The new theory made testable predictions, the most spectacular of which was the existence of a new kind of light made up, not of ordinary photons, but of mysterious particles called Z. In 1983, in a series of high-energy collision experiments in a subatomic particle accelerator near Geneva, Z particles were finally produced, and the unified theory was triumphantly confirmed.

By that time, theorists had already forged ahead and formulated a much more ambitious theory in which the other type of nuclear force (the strong one) is unified with the electromagnetic and weak force. Parallel work on gravity had begun to show how it too could be merged with the other forces in a unified theory. Physicists believe there are only these four fundamental forces of nature, and so the way lies open for a completely integrated theory in which all the forces are encompassed within a single descriptive scheme. The unified field theory, sought for so many decades, seems at last to be at hand.

In their attempts to amalgamate the four forces of nature into a common superforce, physicists have stumbled across some rich bonuses. The modern theory of forces has grown out of quantum physics, in which force fields act by conveying 'messenger' particles. Because all matter is composed of particles as well, quantum physics provides a common description of force and matter. Indeed, it is impossible to untangle the nature of forces from the microscopic structure of matter: particles act on other particles (and themselves) by the exchange of still more particles. It follows that a unified theory of the forces is also a unified theory of matter. The bewildering array of particle species catalogued by experimenters over the past fifty years are no longer a meaningless jumble; they can be ordered into a systematic pattern.

Fundamental to the unification programme is the concept of symmetry. At its most basic, symmetry is present whenever connecting links exist between different parts of an object or system. If subatomic particles with closely related properties are grouped into families, the emergent patterns suggest deep symmetries at work. Mathematical analysis of the forces that shape matter also reveals hidden symmetries of a subtle and abstract nature. Building upon this, physicists have discovered that forces can be understood in a curious way: they are simply nature's attempt to maintain various abstract symmetries in the world.

It is from these insights into the relationship between force fields, particles and symmetry that there has come perhaps the most remarkable conjecture of all – that we live in an eleven-dimensional universe. According to this theory, the three-dimensional space of our perceptions is augmented by seven unseen space dimensions which, together with time, make up eleven dimensions in all. Although the extra seven dimensions are invisible to us, they still manifest their existence as *forces*. What we think of as, say, an electromagnetic force is really an unseen space dimension at work. The geometry of the seven extra dimensions reflects the symmetries inherent in the forces. It follows from this work that there really are no force fields at all, only empty eleven-dimensional spacetime curled into patterns. The world, it seems, can be built more or less out of structured nothingness. Force and matter are manifestations of space and time. If true, it is a connection of the deepest significance.

From these exhilarating advances in our understanding of the basic forces that build the physical world comes a realization that the essential structure of the universe today was laid down at the earliest cosmic epochs, when the universe was far less than one second old. Astronomers now accept that the cosmos came into being abruptly in a 'big bang', a violent, explosive outburst in which the physical conditions far exceeded the most extreme limits of temperature and compression found in the universe today. For a fleeting instant space was filled with exotic forms of matter controlled by forces that have remained suppressed ever since. It was in that first brief flash of existence that the superforce reigned supreme.

In the beginning the universe was a featureless ferment of quantum energy, a state of exceptionally high symmetry. Indeed, the initial state of the universe could well have been the simplest possible. It was only as the universe rapidly expanded and cooled that the familiar structures in the world 'froze out' of the primeval furnace. One by one the four fundamental forces separated out from the superforce. Step by step the particles which go to build all the matter in the world acquired their present identities. It was also at this very early stage that the beginnings of galaxies were generated. One might say that the highly ordered and intricate cosmos we see today 'congealed' from the structureless uniformity of the big bang. All the fundamental structure around us is a relic or fossil from that initial phase. The more primitive the object, the earlier the epoch at which it was forged in the primeval furnace.

The greatest cosmic mystery has always been what caused the big bang. Until now, only metaphysical answers to this question have been forthcoming. Today we can glimpse a properly scientific explanation for the big bang based on the activities of the superforce. According to these very latest ideas, the universe erupted into physical existence spontaneously, out of literally nothing. Even space and time came into being at this moment. The secret of this cosmic event without a cause is quantum physics, a topic which will be discussed at length in the coming chapters.

Once it had come into existence, the universe evolved exceedingly rapidly under the control of the superforce. Some theorists believe that the currently observed large-scale structure of the universe was established in the first 10^{-32} s and that this speedy development of cosmic order included the transformation from ten space dimensions to the three that survive today. It was also during this era that the universe could have become caught in a 'cosmic bootstrap', which enabled it to self-generate huge quantities of energy from nothing. If so, then it was from this primeval energy that originated all the matter which went to build the cosmos, and all the energy which still powers the universe today.

Scientists themselves divide into two camps. There are those who believe that, in principle, science can explain the universe in its entirety. The others insist that there is an irreducible supernatural or metaphysical element to existence that cannot be tackled by rational enquiry. The scientific optimists, if we may call them that, are not so bold as to claim that we will ever achieve a complete working knowledge of every detail of the cosmos, but they do maintain that every process and every event conforms strictly with the rule of natural law. Their opponents deny this.

Of all the sciences, physics faces this stark choice most acutely, partly because it is a 'fundamental' science. For it is the job of the physicist to understand the nature of space and time, the basic structure of matter and the operation of the forces that build the objects which collectively we call the universe. It is the ultimate goal of physics to explain what the world is made of, how it is put together, and how it works. If any part of the world, past, present or future, cannot be encompassed in this programme, it is the physicist who is most likely to be alarmed.

Even as recently as the mid-1970s some of the achievements described in this book would have been unthinkable. Most cosmologists held that although physics could explain the development of the universe once it had been created, the ultimate origin of the universe lay beyond the scope of science. In particular, it seemed necessary to suppose that the universe was set up in a very peculiar state initially, in order that it might evolve to the form we now observe. Thus, all important physical structures, all matter and energy, and their large-scale distribution had to be assumed as god-given; they had to be put in 'by hand' as unexplained initial conditions. With the recent breakthrough in understanding, all these features emerge naturally and automatically as a consequence of the laws of physics. The initial conditions, inasmuch as we may even make sense of the concept in a quantum context, exercise no influence on the structure of the universe which emerges subsequently. Thus the universe is seen to be a product of *law* rather than *chance*.

The fact that the present nature of the universe was bound to have developed from the big bang origin – it is written into the laws of physics – strongly suggests that these laws are not themselves accidental or haphazard, but contain an element of design. Despite the decline of traditional religion, ordinary men and women continue to search for a meaning behind existence. The new physics and the new cosmology reveal that our ordered universe is far more than a gigantic accident. I believe that a study of the recent revolution in these subjects is a source of great inspiration in the search for the meaning of life.

As always in science, theories and models are tentative and liable to be proven wrong as new discoveries are made. Many of the topics discussed in this book are at the very forefront of research, and I do not doubt that future developments will result in a major reappraisal of their significance. Caution is therefore urged concerning some of the detailed results to be presented. Having said this, I do not believe that future developments will challenge the essential theme of the book, that for the first time in history we have within our grasp a complete scientific theory of the whole universe in which no physical object or system lies outside a small set of basic scientific principles. The theory of the universe to be expounded here may turn out to be wrong, but at least we can now glimpse what a complete theory of all existence is like. We can see how such a theory is possible. We can at last comprehend a universe free of all supernatural input, a universe that is completely the product of natural laws accessible to science, yet which possesses a unity and harmony that manifests insistently a strong sense of purpose.

Where are we?

One of my earliest childhood memories is of asking my father where the universe ends. 'How can it end?', he replied. 'If space had an end, something else would have to lie beyond it.' It was my first encounter with the concept of infinity, and I still remember the mixture of puzzlement, awe, and fascination. As it turns out, the issue is not quite as straightforward as my father led me to believe.

To understand about the limits of the universe, you first have to know about our location within it. Planet Earth, together with its eight sister planets which orbit the sun, make up the solar system. The sun is a typical star, and the other stars we see by glancing at the night sky are relatively nearby suns (perhaps a little larger and brighter than our sun), probably with their own planetary systems too. The stars are not distributed haphazardly in space, but organized into a colossal wheel-shaped structure called the galaxy. The broad band of light known as the Milky Way is revealed in a telescope as a huge collection of stars, gas, and dust, and represents the brighter portion of our galaxy. The Milky Way appears as a band of light because the galaxy is disc-shaped. We see most of the stars when we look in the plane of the disc. The sun is located in this disc about two-thirds of the distance from the centre. The galaxy does not have an abrupt edge; the whole structure is embedded in a large distended halo of widely spaced stars.

Looking beyond the confines of the galaxy, it is possible to see other galaxies, very similar in form to our own, scattered around us in a cluster. Typical of these galaxies is Andromeda, which is just visible to the naked eye as a hazy patch of light. This local group is in turn part of a larger aggregation of galactic clusters, and so on. Modern telescopes reveal a universe full of clusters of galaxies, thousands and thousands of millions of them, distributed evenly throughout space. Galaxies are the building blocks of the cosmos.

Distances in astronomy are awesome in their magnitude. Expressing them in miles or kilometres, one soon gets lost in a maze of zeros. A more convenient unit is the light-year, being the distance travelled by light (the swiftest entity) in one year. A light-year is about 9½ million million kilometres (six million million miles), but is more readily visualized by remembering that light takes only 8.5 minutes to reach Earth from the sun which is 150 million kilometres (93 million miles) away. The moon is about one light-second from Earth. In these units, the solar system is a few light-hours across and the nearest star is a little over four light-years away. The galaxy is roughly 100 000 light-years in radius and in this vast space are contained at least 100 000 million stars. Distances to other galaxies are measured in millions of light-years. Nearby Andromeda is about 2.5 million light-years away, and the world's largest telescopes can detect galaxies as far away as 10 000 million light-years.

This picture of the universe has been constructed in only relatively recent times. All ancient cultures took it for granted that the Earth was situated at the centre of the cosmos. Though astronomy was well developed in many primitive societies, a proper understanding of the nature of the stars and the large-scale structure of the universe had to await the modern scientific era.

In pre-scientific Europe, cosmological ideas tended to reflect the work of the early Greek philosophers. Pythagoras, in the sixth century B.C., had conceived of a spherical Earth at the centre of a spherical universe. Cosmic bodies were considered divine and their circular revolutionary movements were regulated to geometrical perfection. Over the centuries the Greeks developed this basic theme, culminating in the complex model of Claudius Ptolemy in the second century A.D. Ptolemy's universe involved a collection of interlocking rotating spheres designed to reconstruct the complicated motions of the moon and planets.

These early model universes were usually finite in size, but there was much anxiety about the nature of the cosmic edge. The Roman poet Lucretius drew attention to the issue by asking what would happen if someone were to make his way to the 'uttermost boundary' and hurl a spear. Would its path be blocked? In some models the answer was yes, since the cosmos was considered to be bounded by some sort of wall or impenetrable surface, a strange idea which survived right up to the time of Kepler in the seventeenth century.

In contrast to a sharp boundary, Aristotle favoured a gradual fading away of the physical realm into the world of spirits and ethereal substances. Even today some people cling to this idea, envisaging a 'heaven' beyond the sky, and much of our superstition and religious symbolism is based upon similar notions. Indeed, the word 'celestial' still refers to both the astronomical and the spiritual. An alternative cosmological tradition was that of 'the void'. In this model, the material universe was regarded as finite, but its outer limits did not mark the boundary of all existence. Instead, empty space lay beyond, stretching away to infinity. Whatever the nature of the cosmic edge, however, the Earth was always considered to be at the centre of the universe.

These ideas collapsed in the Middle Ages, when Nicolas Copernicus proclaimed that the sun, rather than the Earth, lay at the centre of the universe, and that the planets revolved around this centre. Copernicus' model universe was still finite in size and possessed an outer edge – a sphere containing the fixed stars. Shortly afterwards Thomas Digges proposed abolishing Copernicus' outer edge in favour of a scheme in which the stars were scattered throughout unbounded space. The concept of an infinite universe had been espoused by Lucretius and the so-called Atomist School a thousand years previously, but the mystical and religious aspects of the infinite frequently got in the way and clouded the issue. Giordano Bruno, for example, was burned by the Church for suggesting that there was an infinity of worlds.

The growth of scientific astronomy, particularly the development of large telescopes and the invention of the spectroscope, dramatically changed mankind's conception of the universe. The Milky Way was seen to be an 'island universe', with a discrete identity. Even at the turn of the last century there was still controversy over whether the Milky Way galaxy sat alone in an infinite void, or whether there were other 'island universes' beyond ours. To some astronomers, at least, it seemed conceivable that one could travel far into space and look down on all of creation, the whole universe of stars being concentrated in one region and unbounded emptiness lying beyond.

The true nature of the universe was not fully decided until the 1920s, following the work of the American astronomers Harlow Shapley and Edwin Hubble. They finally established that many of the so-called nebulae – fuzzy patches of light long familiar to astronomers – were other galaxies lying beyond our own. As far as our telescopes can probe into space there are galaxies. No evidence is found for a fall-off in their density, and there is no suggestion of an edge or limit to the collection of galaxies. Cosmologists prefer to believe that there is no edge to the assemblage of galaxies, and that wherever there is space there are galaxies. In spite of this many people (including some scientists) still envisage the universe as a cluster of galaxies surrounded by an infinite void. Popular articles often refer to 'the edge of the universe', implying some outer limit beyond which there is only emptiness. The official position, however, is that there is no cosmic edge and no cosmic centre. The universe is not a collection of galaxies contained within space; rather, space is contained within the universe.

Paradoxically, it is not necessary to assume that an edgeless universe is infinite in volume with an infinity of galaxies. One of the curiosities of modern cosmology is that the cosmos could be finite and yet without boundaries. If this seems like a contradiction, think about the properties of a circle. In one sense a circle 'goes on forever'. It has no boundary or edge, nor does it have a centre (at least not one that lies on the circle). In spite of this a circle is still finite. We might say that a circle is a line which curves around and joins up with itself. It is possible to generalize this idea to three dimensions and to imagine the universe curving around and joining up to form a finite space without boundary. Many people have trouble trying to visualize a closed and finite universe: they want always to imagine that something lies outside it. Nevertheless, this concept makes logical sense and can be given a proper mathematical description. There is no agreement among cosmologists, however, of whether the universe really is like this.

If there is no outer limit to the realm of the galaxies, the question 'Where are we?' loses much of its meaning. Space itself has no landmarks, and even in the most remote regions the large-scale aspect of the cosmos is much the same as in our own galactic neighbourhood. On a small scale the concept of 'Where?' makes good sense, because we can measure our location relative to some conspicuous nearby object, such as the sun or the centre of the galaxy. But within the universe as a whole there is no privileged place from which objects can gauge their location. It is rather like standing on an infinite chequerboard: you can give meaning to how far you are from the nearest corner of a square, but your overall position on the board is a meaningless concept.

When are we?

Though we can make no general sense of the question 'Where?', cosmologists often talk about the age of the universe. The issue is much the same for time as it is for space, and there is a long history of argument and muddle concerning this topic too. Plato taught that the world of God's creation was perfect and therefore unchanging in its broad features. He maintained that, though day by day we notice change, over the aeons things remain more or less the same. If Plato's belief were correct the world could not have been created at some moment in time, but would have endured for all eternity. The question 'When are we?' would then be meaningless because time would have no beginning.

An alternative tradition is that of a created world in which the universe is of finite age and undergoes continual irreversible change. Obviously, if the original creation were perfect it would not then remain so, though it could presumably start out marred and evolve towards (or strive for) perfection.

Mythical accounts of genesis are legion, and usually extremely fanciful. The scientific version of the creation has been developed in detail only in modern times. Its origins go back to Hubble's work on extragalactic objects in the 1920s. It was as a result of his careful research on the spectra of distant galaxies that Hubble made a momentous discovery which was to lay the foundation for all modern scientific cosmology. He found from the way in which their light is distorted – or 'red-shifted' – that the galaxies are receding from us at very great speed. A systematic study of the pattern of motion and the way it varied with distance showed that the other galaxies are also moving apart from each other. In fact, the entire universe is everywhere in a state of expansion.

The topic of the expanding universe can also lead to visualization troubles, and it often exacerbates the confusion about 'Where?' There is a great temptation to think of the expansion as the explosion of a concentrated lump of material, with the fragments rushing away into a limitless pre-existing void. This erroneous image pictures the size of the universe growing continually as the outermost members of the collection of fragments recede further and further into the void. However, we have seen that this general picture of the universe is badly misconceived, because it assumes the existence of a cosmic edge. A more accurate picture is one in which the space *between* the galaxies expands. A helpful analogy here is that of a balloon being slowly inflated. Imagine the balloon's surface covered with dots which represent the galaxies. As the balloon expands so the rubber stretches and the dots find themselves farther and farther from their neighbours (*see* Figure 1). Notice that the dots themselves do not move towards or away from anywhere actually within the surface. The separation of the dots is brought about because the surface itself expands.

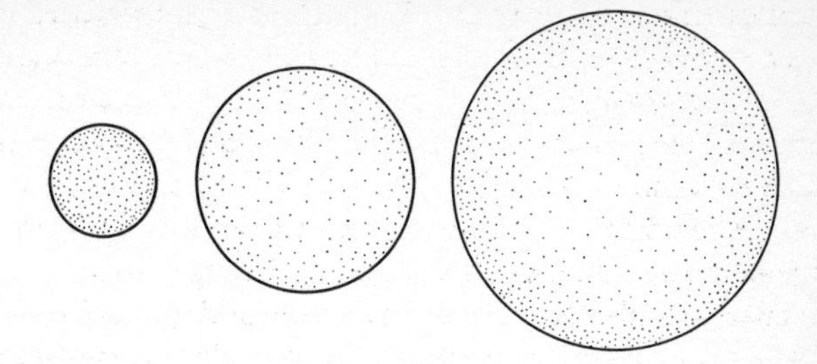

Figure 1. The expanding universe is analogous to an inflating balloon. The dots represent galaxies, and are scattered more or less evenly across the surface. As the balloon swells, so the space between the 'galaxies' stretches. From the viewpoint of any given dot it will appear as if the neighbouring dots are receding, but the dots themselves do not move within the surface. The assemblage of 'galaxies' is *not* expanding away from a point in space. Of course, the two-dimensional surface of the balloon is only an analogy to three-dimensional space; in the real universe there is no physical region corresponding to the interior or exterior of the balloon.

The expanding universe is rather like a three-dimensional version of the expanding balloon. It is therefore wrong to think of the galaxies as rushing *through* space, away from a common centre of expansion. It is the effect of the space between the galaxies swelling or stretching which carries the galaxies away from each other. The ability of space to stretch is a consequence of Einstein's general theory of relativity which will be explained in the forthcoming chapters. The fact that we see the distant galaxies rushing away from us does not imply that we are at the centre of the expanding universe, any more than a given dot in the balloon can be considered to be at the centre of the balloon's surface. (The *surface* does not have a centre.) Therefore the universe does not expand *into* anything; it simply grows in scale everywhere.

If the universe is swelling it must have been shrunken in the past, and extrapolating backwards in time one can deduce that about 15 000 million years ago the cosmic material was highly compressed. This leads to the *big bang* theory of the origin of the universe, which proposes that the entire cosmos came into existence in a huge explosion.

According to the modern version of this theory, the early stages of the big bang were characterized by extreme heat and density, so that none of the structures we now observe in the universe, including atoms, could have existed. Important confirmation of this scenario was obtained in 1965 when two telecommunications scientists working for the Bell Telephone Laboratory stumbled across a mysterious source of radiation from space. Physicists and astronomers were quick to identify this cosmic background radiation as a relic of the primeval heat, a fading glow from the fiery outburst that marked the creation event 15 000 million years ago.

The nature of the big bang is frequently misconceived for it is often presented as the explosion of a lump of material in a pre-existing void. But as we have seen, there *is* no space outside the universe. Rather, it is more accurate to envisage the big bang as an event in which space itself came into being. In fact, the scientific picture of the creation is, in this respect, more profound than the biblical, for it represents the origin not only of matter, but of space too. Space came out of the big bang, and not the other way round. The big bang, then, was not an event which occurred within the universe; it was the coming-into-being of the universe, in its entirety, from literally nothing.

Another important feature of the big bang concerns time. Many cosmologists believe that time did not exist before the big bang, i.e. that there was no 'before'. One of the lessons of the new physics is that space and time are not simply there, they form part of the physical universe. Therefore, if the big bang marked the origin of the physical universe, space and time came into existence only then. Actually, identifying the creation of the universe with the beginning of time is not a modern idea. In the fourth century Saint Augustine wrote, 'The world was created with time and not in time.'

The abrupt appearance of the universe in a big bang means that it makes sense to ask 'When are we?' All cosmic epochs can be referred to this single, profound event which occurred about 15 000 million years ago. We can chart the history of the universe as it evolves over the aeons, and gauge all dates from this absolute zero of time.

What are we?

The simple answer to this next question is that we are matter. But what is matter and how did it come to exist? Such is the stupendous range of shape, colour, density, and texture of material things, it might appear a hopeless task to attempt to understand the nature of matter. Yet two and a half millenia ago the Greek philosophers laid the foundations for this understanding when they set about explaining the complexity of the world by reducing it to the interplay of primary constituents. In the sixth century

B.C. Thales proposed that the primary element of all things was water, but later thinkers conceived of four terrestrial elements, earth, air, fire, and water. These four elements, it was thought, were conserved – their total quantity remained unchanged – but they could combine together in a great variety of shapes and compositions. The heavenly bodies were composed of a fifth substance, called ether or quintessence. Greek philosophers made an important step in that they at least rejected magical argument and observation – the essence of the scientific method. Anaxagoras (500–428 B.C.) greatly improved on the earlier theories by conceiving of an infinite universe populated by an infinite number of particles, or 'atoms'. Furthermore, Anaxagoras proposed that the heavens were made of the same substances as the Earth, a heresy that almost cost him his life. Leucippus also developed the atomic theory of matter, which was then elaborated by his student Democritus, but the theory fell into disfavour, being rejected by great philosophers such as Aristotle, Plato, and Socrates. However, atomist ideas were later revived by Epicurus (341–270 B.C.).

The essential feature of atomism was that the world is held to consist of just two things, indestructible atoms and the void. The atoms come in a number of shapes, and can lock together in many different ways to form composite systems. Atoms are indivisible and move freely through the void. They are in a state of continual activity, for ever colliding and uniting into new forms, subject always to the rational laws of cause and effect.

For centuries the atomic theory of matter remained a mere speculation, the atoms being too small to permit direct observation. Competing ideas of the continuum, in which matter is infinitely divisible and contains no void, remained alive even until the twentieth century. With the rise of systematic chemistry, the atomic theory entered modern scientific thinking. The English chemist John Dalton (1766–1844) offered evidence that atoms have different weights and combine in certain fixed proportions to form compounds, but direct physical evidence for atoms was still lacking. Only by the end of the nineteenth century, with the discovery of the electron and radioactivity, were atoms finally discerned. It soon became clear that there are many different types of atom, each corresponding to the modern version of a chemical element. Today about ninety naturally occurring elements have been identified on Earth, and a dozen or more produced artificially.

In 1909 the New Zealand physicist Ernest Rutherford established the basic architecture of the atom. Rutherford bombarded atoms with alpha particles from radioactive emissions, and determined from the pattern of their scattering that atoms were not like hard lumps of indivisible matter as some physicists had believed, but composite bodies with most of their mass concentrated in a central nucleus which is surrounded by a cloud of lighter, mobile electrons (*see* Figure 2). This structure is reminiscent of a planetary system. The force of attraction that binds the electrons in their orbits is produced by the electric charge on the nucleus.

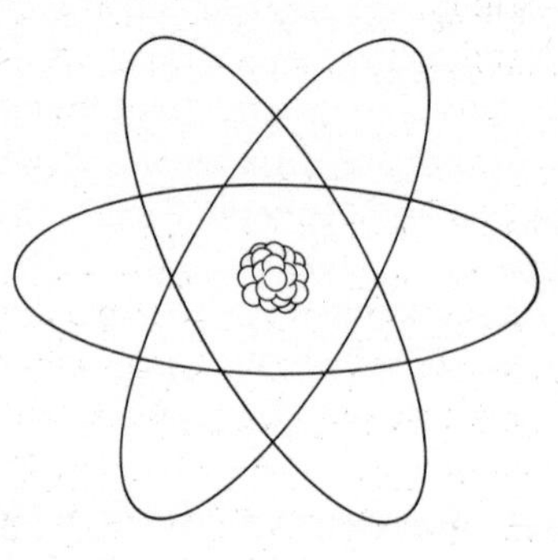

Figure 2. Schematic representation of an atom. The central nucleus, consisting of a ball of tightly bound protons and neutrons, is surrounded by a cloud of orbiting electrons. Most of the mass is contained in the nucleus. Because of quantum effects, the electron orbits are not really well-defined paths as depicted here.

It was not until the early 1930s that the nature of the nucleus was properly understood. This, too, turned out to be a composite system consisting of a ball of protons and electrically neutral particles called neutrons. Today it is believed that both protons and neutrons are in turn composed of still smaller entities known as *quarks*. Many physicists believe that electrons and quarks are truly elementary particles in the Greek sense. They appear to have no internal structure whatever, and together they build up all known forms of ordinary matter.

Evidently matter is a hierarchy of structure. Quarks are used to build protons and neutrons, which in turn build the nuclei which go into atoms. Atoms combine to form molecules or crystals. These basic materials then make up all the solid objects around us. Continuing upwards in scale, we come to the planetary systems, star clusters and eventually galaxies, and even the galaxies in turn cluster into larger groups and supergroups. Human beings come somewhere in the middle of this hierarchy. Measure for measure, we are to an atom what a star is to us.

It is well known that some elements, such as oxygen and iron, are rather plentiful, whereas others, such as uranium and gold, are so rare that people occasionally go to war in order to secure a supply of them. If the relative abundances of the elements in the universe as a whole are estimated, a striking pattern emerges. About 90 per cent of cosmic material is hydrogen, the lightest and simplest substance. Hydrogen atoms consist of a single proton and a single electron. Of the remaining 10 per cent, most is helium, the next simplest element. Helium nuclei contain two protons and two neutrons. The remaining elements constitute less than 1 per cent of the total. With the exception of iron, the general trend is that the heavier elements such as gold, lead and uranium are far less abundant than the light elements such as carbon, nitrogen, and oxygen.

This pattern of varying abundance is very suggestive. Heavy nuclei contain many protons and neutrons; light nuclei contain few. If light nuclei could be fused, heavier nuclei would be produced. It is tempting, therefore, to suppose that the universe began with only the simplest element, hydrogen, and that the heavier elements have been built up, step by step, in successive stages of nuclear fusion. This theory immediately explains why heavy nuclei are rare. Fusion requires very high temperatures to overcome the electric repulsion between nuclei. The more protons a nucleus contains, the greater the repulsion and the harder it is to add still more protons in a fusion reaction.

Explaining how the chemical elements came into existence is only a partial solution of the puzzle concerning the origin of matter. One can still ask how the protons, neutrons and electrons that go to make up these elements came to exist in the first place.

Scientists have long known that matter is not permanent, but can be created and destroyed. If enough energy is concentrated, new particles of matter will come into existence. We can envisage matter as a form of locked-up energy. The fact that energy can be converted into matter suggests that the universe began without any matter and that all the material we now see was generated from the energy of the big bang. This attractive theory faces a serious obstacle, however. The creation of matter in the laboratory is now routine, but every newly created particle is acompanied by a sort of 'negative image' partner, known as an *antiparticle*. For example, an electron (which carries a negative electric charge) is always created along with an antielectron, more often called a positron, which has the same mass as an electron but an opposite (positive) electric charge. Similarly, every created proton is accompanied by an antiproton.

Collectively, antiparticles are known as antimatter.

When a particle encounters an antiparticle, mutual annihilation results, with the release of all the locked-up energy. Obviously, a mixture of matter and antimatter is violently unstable. For this reason it seems unlikely that more than a tiny fraction of the universe is made of antimatter. The problem is then to understand how matter came to exist without an equivalent quantity of antimatter. We shall see how recent discoveries suggest a solution to this problem.

The creation of matter from energy is not restricted to the familiar particles such as electrons, protons, and neutrons. Other, more exotic forms of matter can also be produced. In fact, hundreds of different subatomic fragments have been created in the laboratory by colliding fast-moving particles, using accelerators. These other particles are all unstable and rapidly decay into more familiar forms. Being so short-lived, they play no direct role in the universe.

How are we put together?

If there were no forces, particles of matter would simply move about independently, unaware of each other's existence. The presence of forces enables particles to recognize and respond to other particles, and therefore display collective behaviour.

When an engineer talks of forces, he usually has in mind something which pushes or pulls, such as a rope or wire. This sort of force we can readily visualize, and we can easily understand from direct experience how such forces can move matter about. There are, however, other manifestations of forces that are less familiar, such as the radioactive decay of an atomic nucleus or the explosion of a star. Because all matter is made up of particles we must ultimately turn to particle physics for an explanation of forces. When this is done, it is found that all forces, whatever their large-scale manifestations, can be reduced to just four basic varieties: gravity, electromagnetism, and two types of nuclear force. We shall see in later chapters how the forces are actually communicated from one particle to another. We shall also see that forces and particles are closely related, so that we cannot understand one without understanding the other.

As the scale of size increases, so the relative importance of the four forces changes. At the level of quarks and nuclei, the two nuclear forces dominate. The strong nuclear force is responsible for binding the quarks into protons and neutrons, and for holding atomic nuclei together. At the level of atoms, electromagnetism is the dominant force, binding the electrons to the nuclei and enabling atoms to combine together into molecules. Most of the 'everyday' forces, such as the tension in a wire or the push of one object against another, are examples of the large-scale action of electromagnetic forces. When it comes to astronomical systems, gravity is the dominant force. Each force therefore comes into its own at a certain scale of size, and each has an important role to play in shaping the features of the physical world.

In recent years physicists have come to wonder about the relationship between the four forces that together control the universe. Is there a connection? Are the four forces merely four different manifestations of a single underlying *superforce*? If this superforce exists, it will ultimately account for all the activity in the universe, from the creation of subatomic particles to the collapse of a star. If we unlock this superforce it would give us power beyond all imagination. It might even explain how the universe came to exist in the first place.

2
The New Physics and the Collapse of Common Sense

Warning: physics can expand your mind

'Science is nothing but trained and organized common sense.' Thus wrote the great nineteenth century biologist T. H. Huxley. In Huxley's day this was probably correct. Although nineteenth century science spanned a wide range of subjects, all its concepts were firmly rooted in the commonsense world of daily experience.

Physics had scored many successes by the close of the century. Electricity and magnetism were well understood, radio waves had been discovered, and the atomic theory of matter was put on a firm foundation. But although these topics stretched science beyond the realm of direct human perception, they were still formulated as simple extensions of familiar objects and ideas. Atoms were regarded as merely scaled-down versions of snooker balls. Electromagnetic fields were conceived as stresses in an ephemeral medium called the ether, while light waves were thought to be vibrations of the ether. Thus, although atoms were too small to be discerned individually, and the mysterious ether was both invisible and intangible, nevertheless such entities could readily be visualized by analogy with well-known objects. Moreover, the laws which governed these unseen constructions were assumed to have the same form as those which had been applied successfully to more concrete and familiar physical systems.

Then the new physics arrived. The dawn of a new century heralded an explosion of ideas which shattered the cosy notions of reality that had endured for centuries. Many cherished beliefs and unquestioned assumptions were swept away. The world was revealed as a weird and uncertain place, common sense as an unreliable guide. Physicists were forced to rebuild their model of reality, incorporating features that had no direct counterpart in human experience. Abstract and alien concepts, of which only mathematics could give a proper description, were introduced to accommodate the flood of new discoveries.

It was a time for revolutions; not one revolution but two, coming directly on top of each other. First there was the quantum theory, which provided novel insights into the bizarre workings of the microworld, and then there was the theory of relativity, which cast space and time into the melting pot. The old world view of a rational and mechanistic universe, ordered by rigid laws of cause and effect, collapsed into oblivion, to be replaced by a mystical world of paradox and surrealism.

The first casualty of the twin revolutions was intuition. The nineteenth century physicist could get a good mental picture of his subject matter, but quantum and relativity physics demanded unprecedented mental gymnastics. Some phenomena seemed so hard to imagine that even professional physicists balked. Max Planck, for example, who initiated the quantum theory, never really came to accept its peculiarities in full, while Einstein found it so preposterous he resisted it to his dying day.

The new physics continues to provide novel insights into the workings of the universe, and each new generation of students finds the ideas involved strange or even illogical. A well-known English university used to display a notice above the entrance to its Physics Department building which cautioned, 'Warning: physics can expand your mind.'

Take, for example, the world of subatomic quantum particles, where intuition fails completely and nature seems to play tricks on us. One of these is the barrier trick. Imagine throwing a stone at a window. If the stone is moving slowly it will bounce back, leaving the window intact. With more energy, the stone will shatter the window and pass through. A similar exercise can be performed in the atomic world, where the role of the stone is played by a particle such as an electron, and the window is some sort of fragile barrier, such as can be provided by an array of atoms or an electrical voltage. Often the electron will behave in the same way as the stone, bouncing back when it approaches the barrier slowly and breaking through when it possesses more energy. But sometimes this simple rule is flagrantly violated. The electron is seen bouncing back from the barrier even when it has more than enough energy to force a passage.

Even more bizarre are the occasions when an electron which does not have enough energy to break the barrier nevertheless appears miraculously on the remote side. Imagine tossing a pebble lightly at a window, only to find it penetrate the glass and appear on the far side, leaving the window intact! Yet this piece of trickery is precisely what electrons are seen to do. In effect, they 'tunnel' through an insurmountable barrier. Another trick can occur if an electron is approaching a chasm into which it is about to fall. It may then abruptly reverse direction just as it reaches the edge of the pit. Nor is this wild behaviour at all predictable. Sometimes the electron will bounce back; at other times it will fall into the pit.

These weird phenomena make it seem almost as though the electron can sense its surroundings. When it reaches a barrier it appears to 'see' beyond it and reason, 'The barrier is only thin, so I'll disappear and rematerialize on the far side.' Though the idea that an electron can be here at one moment and there the next seems utterly alien, this is just what happens. In fact, in some ways electrons behave as though they are in many different places at once. It is important to realize that these outlandish antics are not merely a bit of speculative science. The 'tunnel effect', for instance, is exploited in a number of commercial micro-electronic devices, such as the tunnel diode. Indeed, even the ordinary flow of electricity in a copper wire has an element of tunnelling to it.

Many of the oddities of electrons can be traced to the fact that in some respects they behave like waves. In fact, it is possible to demonstrate the undulations of electron waves in several controlled experiments. The idea that something can be both a wave and a particle defies imagination, but the existence of this wave-particle 'duality' is not in doubt. It also happens that what we normally think of as a wave can take on particle aspects in the microworld. Light waves, for example, behave like a stream of particles in the way that they knock electrons out of metal surfaces (the photoelectric effect). Particles of light are known as *photons*, and physicists place them alongside electrons and quarks in the list of fundamental particles. It is impossible to visualize a wave-particle, so don't try. There is nothing in the everyday world that remotely resembles such a monstrosity. If we did come across a wave-particle we should not be able to make sense of it.

A lot of the difficulty people have in understanding modern physics is due to their making futile attempts to force-fit the abstract concepts encountered into an 'everyday' framework of commonsense ideas. It seems that people have a deep psychological need to reduce all reality to simple, easily digested images. When something, such as a wave-particle, crops up that has no counterpart in direct experience, bewilderment, or even outright scepticism, results. Students of physics may feel that they cannot possibly have understood correctly because they have no simple mental image of what is going on. I frequently receive letters or manuscripts from amateur scientists in which, say, a new theory of particle physics is concocted based on commonsense notions. The motivation, according to the authors concerned, is that professional physicists must have got it wrong if they, the authors, cannot understand the concepts involved. No deep principle of nature, they declare, can ever be abstract and unfamiliar. Curiously, nobody seems to denounce abstract art in such vituperative terms.

Electrons are not the only particles subject to the caprice of quantum phenomena. Their properties are shared by all microscopic particles of matter, including quarks. The effects described above all occur at relatively low energy. Even more peculiar are some of the high-energy effects, such as the abrupt appearance of a particle where none existed before, or the decay of an unstable particle into a shower of others. There are even 'Jekyll and Hyde' particles in which two separate entities appear to be fused into a hybrid structure of schizophrenic identity.

Among the strangest of the subatomic particles are the neutrinos. These ghostly objects probably have no mass and travel at the speed of light. They carry no electric charge and are almost completely oblivious of solid matter. So insubstantial are neutrinos that they can easily pass right through the earth and could even penetrate several light-years' thickness of solid lead! Countless millions of them are passing through you as you read these words. Neutrinos come close to being pure nothing, except for a vital property called spin. Sometimes they are depicted as literally revolving, like the Earth spinning on its axis, but in fact the analogy is unsound. The spin possessed by a neutrino has some decidedly odd features, as we shall see.

Astrophysics is another subject where down-to-earth, commonsense notions are shattered. Gravity waves provide a good illustration. These elusive disturbances are nothing other than ripples in space itself, a sort of travelling spacewarp. They are generated whenever material bodies or energy engage in violent motion. Though gravity waves carry energy and momentum, they have no substance as such; they are just undulating emptiness. Equally extraordinary is their intense penetrating power, which even exceeds that of the ephemeral neutrinos. Virtually nothing can stop gravity waves, which makes them exceedingly hard to detect, of course, because they just go by and ignore the detector.

Coming to grips with these outlandish notions taxes the imagination to the limit. Systematic progress would be impossible if it were not for mathematics. Abstract formulae have no need of imagination, and can faithfully describe the most bizarre phenomena so long as the equations used are logically consistent. The penetration of physics by advanced mathematics means that most theoretical work ends up as a maze of incomprehensible symbols. Cryptic mathematics, coupled with the strong mystical flavour of the new physics, imbues the subject with a quasi-religious appeal, the professional physicists playing the role of high priests. Undoubtedly, this has a lot to do with the current popularity of the new physics among people of a religious or philosophical persuasion. Yet it must always be remembered that physics is a very practical subject, and in spite of the Alice-in-Wonderland flavour of some of the concepts, much of modern technology depends upon our present understanding of these abstract ideas.

Spacewarps

Among the kaleidoscope of strange images emerging from the new physics, those that belong to the quantum theory and the theory of relativity excite the greatest interest. In its most developed form, the quantum theory is known as quantum mechanics, and in essence it deals with all activity on a microscopic scale. Quantum mechanics provides the foundation for all our understanding of the molecular, atomic, nuclear, and subnuclear realms. The theory of relativity deals with the nature of space, time, and motion. It becomes important when the system of interest is moving close to the speed of light or in an intense gravitational field.

Quantum and relativity physics assault common sense in many ways. Not least among the casualties is our simple notion of geometry. In daily life a metre is a metre. Once it is defined, any particular unit of length is considered fixed and absolute. Few people would ever contemplate the possibility that what is 1 m today might be 2 m tomorrow, or that your metre might be my one-half metre. Yet not only does the theory of relativity demand that distances have no absolute and fixed meaning, it even suggests experiments in which these discrepancies can be checked. If two observers are in relative motion, so the theory goes, they will measure the *same* object to have *different* lengths. And this in spite of the fact that, at rest, both observers agree exactly on the length of the object of interest.

The shrinkage of distances with speed is known as the Lorentz – Fitzgerald contraction effect after George Fitzgerald and Hendrik Lorentz, and is a fundamental result of the theory of relativity. The effect is only important at near the speed of light, but its existence is beyond doubt. The linear particle accelerator at Stanford in California is a straight tube about 3 km long in our frame of reference. The electrons which plunge down it, however, are moving so close to the speed of light that the length of the tube in their frame of reference is scarcely 1 ft in length! The engineers who designed and operate the accelerator must take this shrinkage into account in practical ways.

If relativity makes nonsense of distance, the situation is worse still when quantum mechanics is taken into account, for this even puts paid to the commonsense idea of 'place'. Everybody takes it for granted that all material things must be somewhere. Each subatomic particle that goes to make up, say, your body ought to have a definite place or location. After all, how can a particle truly exist if it isn't somewhere?

When physicists began to investigate the concept of location in the light of quantum physics, to their profound shock they found that, in general, the very notion is meaningless. The source of all the trouble is a fundamental rule of quantum mechanics, known as Heisenberg's uncertainty principle after the German physicist Werner Heisenberg, who co-founded quantum mechanics in the 1920s. According to this principle, it is impossible to give a well-defined meaning to both the position and the motion of a particle at the same time. We can certainly discuss the speed (strictly, momentum) of a particle such as an electron, and perform an experiment to measure this quantity. The experiment will yield a definite result. A similar strategy can be followed to determine the position of an electron if we so choose. Whenever we look for it, the electron will be found somewhere. What cannot be done – what is in principle an impossibility – is to determine *both* these attributes at the same moment. Whatever our measurement strategy, the very act of looking for the electron's location disrupts, in an unpredictable way, the motion of the particle. Likewise a measurement of its motion smears our knowledge of its position. The two sorts of measurement are simply incompatible.

The fact that we cannot know both the position and the motion of a particle at one and the same time must not be regarded as merely a result of experimental clumsiness, or lack of instrumental resolution: it is inherent in the nature of things. The very notion of an electron-at-a-place is itself quite meaningless should we choose to have knowledge of its momentum instead.

All this makes nonsense of any attempt to picture the atomic world as populated by little balls whirling about. If a particle cannot have a place *and* a motion, there is no way we can sensibly ascribe to it a path through space. It may be that we know at some instant an electron is at point A, while at a later instant it is at point B. But we cannot know how it got from A to B. The idea of a trajectory, or orbit, continuously connecting points of departure and arrival, evaporates. In fact, we have seen how in some very practical devices electrons are apt to 'tunnel' through barriers by disappearing on one side and then suddenly reappearing on the far side of a barrier. This is a typical quantum effect.

The only way to make sense of this erratic behaviour is to suppose that, in getting from A to B, the particle somehow pursues *all possible paths* at once! This bizarre property can readily be demonstrated by adapting a famous experiment first performed by the English physicist Thomas Young in the nineteenth century. Young was interested in demonstrating the wave nature of light, and this he did using a phenomenon known as interference. Interference occurs when two waves overlap each other. If the peaks of one wave coincide with the peaks of the other, reinforcement results, and the wave motion is amplified. On the other hand, if the peaks of one wave arrive aligned with the troughs of the other, cancellation results and the wave disturbance is reduced.

In Young's experiment (illustrated in Figure 3) a small source of light

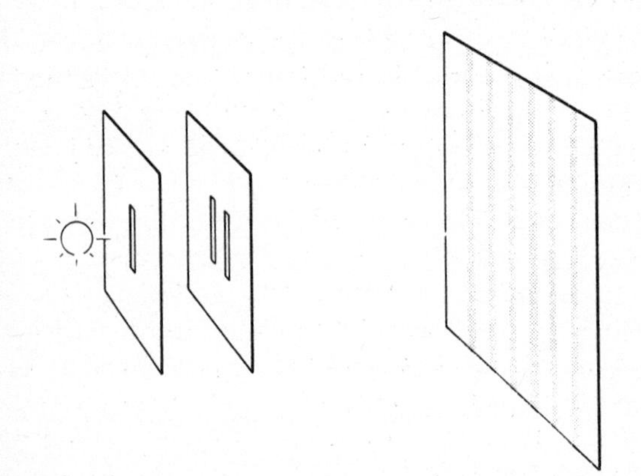

Figure 3. Young's interference experiment. A light source illuminates two parallel slits in an opaque screen. The projected image does not appear as two bands of light, but as a whole series of bright and dark bands, or 'interference fringes'. The experiment graphically illustrates the wave nature of light, but weird overtones develop when the particle (photon) aspect of light is contemplated.

illuminates two nearby slits in an opaque screen. The images of these apertures are projected on to a second screen. Light waves from each slit arrive at the image screen together and interfere. The result depends on whether the two sets of waves arrive in step or out of step. This in turn depends on the angles involved, and will vary from place to place on the screen. The upshot is that a series of bright and dark bands are produced as the light waves alternately re-inforce or cancel.

Weird overtones develop when the quantum nature of light is taken into account. A light quantum, or photon, behaves like a particle to the extent that it arrives at the screen at a definite place. (If the screen is replaced by a photographic plate to record the interference pattern, each photon will chemically alter a single grain of photographic emulsion at a well-localized point.) On the other hand, the interference pattern is clearly dependent on the presence of both slits to produce *two* wave sets that can overlap. If either slit is blocked off, the pattern disappears. Nor is it the case that some photons go through one slit and some through the other, because the pattern will build up, in a speckled sort of way, even if the light is sent through photon by photon. The only explanation is that each photon somehow goes through *both* slits and carries an imprint of their existence when it arrives at the image screen. That imprint serves to direct the photon most probably towards a bright band area (where most photons end up) and away from the dark band regions. In this way, both wave and particle aspects of light co-exist. Although the experiment was performed initially using light, similar considerations apply if electrons or any other quantum 'wave-particles' are used.

The notion of a particle being 'everywhere at once' is impossible to imagine. One can, perhaps, envisage countless 'ghost' particles exploring all the pathways available, only to fuse into a 'real' particle at the point of observation, but even this imagery is inadequate. Only mathematics can encapsulate the subtleties involved.

Being unable to pin down a particle at a place leads to some peculiar effects when more than one particle is involved. If we have a collection of identical particles in a group, and we can't say in any individual case whether a particle is here or there, how can we know which is which? In fact, we don't. The individual identities of the particles become completely blurred.

This ambiguity actually leads to important physical effects. When two atoms bind together to form a molecule, the activity of the electrons around one atom becomes distorted by the presence of the other, producing a force of attraction between the atoms. In part, this force will depend on the fact that a given electron on one atom is indistinguishable from those on the other atom, and because of the fuzziness in their locations, there is nothing to prevent two electrons from swapping places occasionally. In other words, two electrons on different atoms can exchange identitites. Exchange forces are very familiar in chemistry, and have measurable consequences.

All this makes the concept of distance look very shaky. But worse is to come. On closer scrutiny it turns out that, as well as a particle exhibiting fuzziness against the backdrop of space, space itself is fuzzy. It's bad enough that a particle doesn't know where it is, but if places themselves don't know where they are, geometry crumbles beyond comprehension.

The origin of this further bewilderment concerns the peculiar properties of gravity. The theory of relativity, which predicts that distances can stretch and shrink depending on the motion of the observer, was generalized by Einstein in 1915 to include gravitational effects. According to the general theory of relativity, gravity is simply the geometry of empty space and time. But not the sort of geometry we learn at school. Gravity is *warped* or curved spacetime. Not only can space stretch and shrink, it can be bent and distorted. It is just such distortions, according to Einstein's theory, that explain gravity.

Einstein pointed to a number of instances where spacewarps and timewarps could be observed. One of these is the effect of the sun's gravity on the space in its vicinity. During a total eclipse, when the sun's glare is blotted out, it is possible to discern a slight displacement in the charted position of the stars close to the sun in the sky, which are seen by light that passes near to the solar surface (*see* Figure 4). The star beams are visibly bent by the sun's spacewarp.

These tests, and others involving the more powerful gravity of neutron stars, have convinced physicists that gravity really does bend space. One consequence is that space (strictly spacetime) must be considered as elastic, capable of changing its geometrical form. In other words, we can contemplate space *activity*. For example, when a star collapses to form a black hole, the relatively mild spacewarp in its vicinity rapidly escalates to form a grotesquely distorted space prison from which nothing may escape. Another example is the expanding universe discussed in Chapter 1, where the space between the galaxies steadily stretches.

If space can change and move, there are deep implications for quantum physics. Just as Heisenberg's uncertainty principle fuzzes out the activity of particles, so it will fuzz out the activity of space. Mathematical modelling suggests that, on a scale at least twenty powers of ten smaller than an atomic nucleus, space becomes 'foamy' in structure, with violent, spontaneous

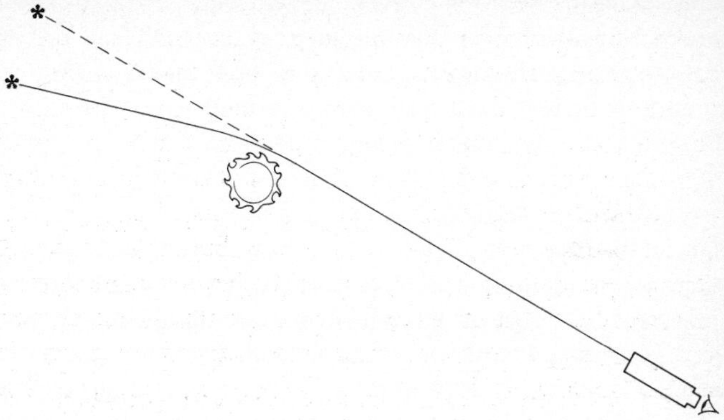

Figure 4. A starbeam passing close to the sun is detectably bent by the solar spacewarp. The effect is to displace slightly the apparent position of the star in the sky.

growth and decay of curvature. In the same way that a particle explores all the pathways of motion available to it, so space on an ultramicroscopic scale explores all its available motions. In the particle case, one way of looking at this was in terms of an army of 'ghost' particles, each following a different pathway. Here we can talk of an infinity of 'ghost' spaces co-existing, each ghost space representing the realization of some particular geometrical form.

This nebulous activity of space implies that the very concept of 'place' breaks down at extremely small distances. The orderly arrangement of points, the smooth continuity of the space of classical geometry, disappears in the froth of spacetime. Instead, we have a mêlée of half-existing ghost spaces all jumbled together. In this chaotic shifting sea, the common sense notion of 'place' fades completely away.

Spin

If places are no longer well-defined in the quantum realm, it comes as no surprise that angles are similarly afflicted. In daily life we take it for granted that objects have an orientation. A vase on the table stands upright, a compass needle points north, a searchlight sweeps the sky. the concept of *direction* is central to our mental model of the world. Without it we could not make sense of external reality.

In the quantum world, however, at the level of atoms and their constituents it is no longer possible to treat direction and orientation naively. An electron orbiting a nucleus cannot be pinned down at any given moment to lie in a particular direction from the nucleus because its position is fuzzy. A beam of photons or other particles cannot be used as a direction-pointer because the particles do not follow well-defined paths; they roam about in an undisciplined way.

In spite of this, there does at first seem to be one promising candidate for an unambiguous definition of direction. It has been mentioned that neutrinos possess a sort of internal rotation or 'spin'. In fact, spin is a property possessed by nearly all subatomic particles, most notably electrons and quarks. It is tempting to picture such a particle, e.g. an electron, as a tiny ball revolving about an axis, like a scaled-down version of the rotating Earth. Obviously, to make sense of such a picture the axis of spin must point along some direction. If this direction can be determined in a measurement, we should have at hand a means to define direction unambiguously, even at the quantum level. Such measurements can be carried out, but here we encounter a most peculiar thing.

Suppose the experimenter sets up his apparatus, and first picks a particular reference direction against which to measure the orientation of the particle's spin. In practice this reference direction could be defined by a magnetic or electric field. The experimenter wishes to know what angle the particle's spin axis makes to the line of the field. He carries out the measurement and he finds to his surprise that the spin happens to point exactly along the direction of the field. The experiment is repeated many times, but the result is always the same. The spin always points along the reference direction chosen. The experimenter suspects some sort of conspiracy and adjusts the angle of his apparatus, but the spin of the particle always follows suit. Try as he may to catch the spin pointing obliquely to the reference direction, the experimenter gets nowhere. He is perplexed by the fact that the particle seems to be reading his mind, because it always anticipates the direction he has freely chosen as his reference.

Frustrated, the experimenter hits upon a devious strategem. He will set up two different reference directions, *A* and *B*, and measure the angle of the spin relative to both. As the spin of the particle cannot possibly point in two directions at once, at least one of the measurements will show the spin at an intermediate angle. Proceeding on this assumption the experimenter makes the first measurement. He is no longer surprised to find the spin pointing along direction *A*. The next measurement he makes very quickly, before something can cause the spin to reorient itself. Direction *B* has been chosen to lie at 25° to direction *A*, and so naturally the experimenter, who has just determined to his satisfaction that the spin points along axis *A*, expects to find the spin pointing at 25° to axis *B*. To his consternation he finds that nature has outmanoeuvred him. Somehow the particle was one jump ahead, and has miraculously re-aligned its spin to coincide precisely with axis *B*. Furious, the experimenter re-measures the angle relative to axis *A*, and, behold, the spin is back at its original angle!

Weird effects like this are now part of established physics, and physicists have long accepted that the spin of a particle will always be found to point along whichever axis is chosen by the experimenter as his reference. It is a property which completely undermines any attempt to make sense of the concept of direction in the quantum domain. It also introduces a bizarre subjective element into the physical world. If the spin of a particle is destined to follow for ever the experimenter's random choice of reference direction, the experimenter's free will somehow intrudes into the microworld. The uncanny slavishness that obliges all spinning particles to adopt the experimenter's definition of angle is suggestive of mind over matter. In Chapter 3 we shall see that these subjective elements of quantum physics demand a complete re-appraisal of the traditional concept of reality and the role of consciousness in the physical universe.

The subject of particle spin contains many other surprises. One of these involves the superficially simple, even trivial, notion of rotation. In daily life we are familiar with the procedure of turning around. Imagine standing in a room facing, say, the door. As you begin to turn, so you face different features of the room until, after a rotation of 180°, you have your back to the door. Continuing on through another 180° you eventually come back to your starting orientation, facing the door, after one complete revolution. The world now looks precisely as it would have done had you not embarked on the rotation. What could be simpler or more obvious?

This elementary act of rotation produces an astonishing result, however, when it comes to subatomic particles. If an electron is passed through a specially shaped magnetic field, its axis of spin can be progressively tipped. Eventually the axis can be rolled right around through 360°. Naturally we would expect, on the basis of common sense, that the electron is now restored to its original configuration. Not so. If the rotated electron's properties are compared with those of an electron left undisturbed, they differ conspicuously. To restore the rotated electron it is necessary to turn its spin axis through a *further* 360°, making *two* complete revolutions in all. There is then no discernible difference between a rotated and an unrotated electron.

What does this mean? Evidently, in the primitive case, a rotation of 720° is necessary to produce a complete revolution, i.e. to restore the world to its original configuration. An elementary particle, such as an electron, perceives the total sweep of 720°. In human beings, and other large objects, this facility is lost, and we cannot distinguish one 360° rotation from the next. In some sense, then, we perceive only half the world that is available to the electron.

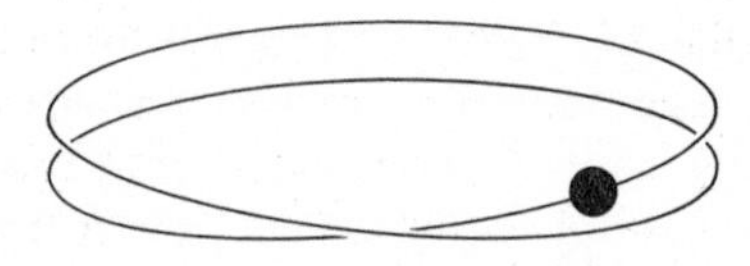

Figure 5. A double loop of wire gives a crude representation of the rotation properties of intrinsic spin. If the bead is slid through 360° it does not return to its starting configuration. This requires a further 360° revolution. From a distance, however, this subtlety is not apparent.

A simple analogy is depicted in Figure 5, which shows a bead threaded on a double loop of wire. From a distance we could not distinguish the two loops, and the wire would appear as a single circle. If the bead is slid around through 360° we should suppose it to have returned to its starting point. On closer scrutiny, however, this is seen not to be so. The bead must be slid a further 360° for it to complete an entire circuit around the loop and properly return to its origin.

The curious 'double-image' view of the world possessed by electrons and other quantum particles, is considered to be a fundamental property of nature. It leads to many unexpected, observable effects. For example, the magnetic field produced by the electron's spin is twice the value that would be generated by spinning an electrically charged ball. In later chapters we shall see how the curious geometrical nature of spin could prove to be the key to unifying physics.

Timewarps

If the new physics plays havoc with our geometrical intuition, it is equally brutal towards the popular notion of time. Common sense leads us to think in terms of *the* time – something universal and absolute, against which all events are gauged. We do not distinguish between my time and your time; there is only *time*. The theory of relativity does not permit this simple-minded state of affairs. Just as space can stretch or shrink depending on the motion of the observer, so too can time. Two events may be judged by one person to be one hour apart, and by another to be one minute.

This is not merely a psychological effect. Time really can be stretched, or 'warped', even in the laboratory, and precision clocks can be used to register timewarps. To produce a timewarp you have to move a clock at very high speed, close to the speed of light. Light travels at 300 000 kilometres per second (300 000 km s^{-1}), which is way beyond the speed of even the swiftest available spacecraft. Nevertheless, such is the accuracy of modern atomic clocks that minute timewarps can be discerned even aboard jet aircraft.

Really spectacular timewarps can be observed using subatomic particles, which are so insubstantial they can be accelerated to very nearly the speed of light. In a recent experiment at the European Centre for Nuclear Research (CERN), for example, it was possible to boost particles called muons so close to the speed of light that their time-scale was stretched over twenty times. Muons are convenient to use because they are unstable, and decay into electrons and other particles after a small fraction of a second. This they do with a fixed half-life, which provides them with an internal clock. In the frame of reference of the muons, decay occurs on average after about two-millionths of a second, but in the laboratory frame this lifetime is greatly stretched.

The stretching of time by motion is one of the effects that some amateur scientists love to hate. It seems to offend their sensibilities more than any other oddity of modern physics. Probably half the papers received by professional journals from private addresses return to the theme of time and relativity, seeking to find a flaw in Einstein's ideas or some contradiction in the theory. They just cannot accept that time is elastic and can stretch or shrink relative to other observers. Particular ingenuity is displayed in trying to shoot down the famous 'twins effect', in which one twin takes a high-speed rocket journey and returns to find his brother ten years older than himself. A phenomenon that professional physicists regard as an amusing curiosity can produce a deep revulsion in others. Perhaps it is because time is a personal experience and some people regard tinkering with time as an assault on their very personality. But like it or not, timewarps are real.

One of the biggest man-made timewarps is produced in a machine at the

Daresbury Laboratory in Cheshire, England. The device, called an electron synchrotron, is designed to accelerate a beam of electrons around a circuit 30 m across, three million times a second. Large magnets deflect the electrons from their natural straight line motion, and with each deflection a burst of electromagnetic radiation, called synchrotron radiation, is produced. The electrons travel to within one-ten-thousandth of 1 per cent of the speed of light, so close that their time-scale gets out of step with ours by a factor of about ten thousand. This discrepancy is exploited by the engineers; indeed, it was the whole point of building the machine in the first place. Although the frequency of the radiation which comes off is only a few kHz (about radio-wave frequency) in the frame of reference of the electrons, the timewarp boosts that frequency by thousands in the laboratory frame. The emerging radiation is actually perceived to be in the ultraviolet or X-ray region of the spectrum. The synchrotron thus uses the timewarp to generate prolific amounts of short-wavelength radiation over a range of frequencies. Such a facility is rare, and has a number of practical applications. At Daresbury, the mystical-sounding timewarp has become a commercial gadget.

The stretching of time goes hand in hand with the shrinking of distance – in fact, the theory of relativity obliges us to link space and time together into a unified *spacetime* – and both warping effects escalate without limit as the speed of light is approached. For this reason it is impossible to break the light barrier and travel at superluminal speed, for to do so would turn spacetime 'inside out', twisting space into time and time into space, enabling objects to travel into the past. The speed of light is therefore regarded as the fastest speed in the universe for the propagation of physical objects or effects.

Timewarps can also be produced by gravity. Time flows a little faster at the top of a building than in the basement, and though the effect is far too small for human beings to notice, special nuclear clocks can detect a timewarp even over the height of a building. Clocks have also been placed aboard high-flying aircraft and rockets to check the effect of gravity on time. There is no doubt about the reality of timewarps; time runs measurably faster in space than on Earth.

The Earth's gravity is modest by astronomical standards, and cosmic objects are known that warp time by huge amounts. At the surface of a neutron star, for instance, where a teaspoonful of neutronic matter weighs more than all the Earth's continents, gravity can be powerful enough to slow time to about half the rate it flows on Earth. If gravity rises much above the neutron-star level then a black hole will result. In this case, the star implodes completely and envelopes itself in an infinite timewarp, as well as a prison of curved space. In a rough sense, time at the surface of a black hole stands completely still relative to our own time-scale.

The fact that time is not fixed and universal, but elastic and flexible, undermines many commonsense beliefs. If my time can get out of step with your time, because of our different motions or gravitational situations, then it makes no general sense to talk about 'the time' or 'now'. the notion of 'this moment' on, say, Mars, is hopelessly ambiguous once one entertains the possibility of fast-moving observers. Likewise, to ask, 'What time is it on a neutron star?', is totally devoid of meaning. Time is purely relative. In our own frame of reference it proceeds at an orderly rate. No matter how we move or change our gravitational experiences it will seem normal to us. But peculiar effects appear when we compare times between two different systems. Then we find that each frame of reference has its own time-scale, and this scale will usually disagree with everybody else's.

Normal is abnormal

The weird effects of quantum physics and relativity on our traditional ideas of space and time imbue the world with a vagueness and subjectivity that belies its everyday normality. Normality is a consequence of the exceedingly limited range of experience with which we are familiar. In our daily lives we never travel at speeds great enough for timewarps and spacewarps to become noticeable, and most of us do not delve into the fuzzy and nebulous realm of the atom. Yet the rational, orderly, commonsense world of experience is a sham. Behind it lies a murky and paradoxical world of shadowy existence and shifting perspectives.

The nebulous surrealism exposed by the new physics is particularly acute when it comes to matter. The solid dependability of, say, a rock, reassures us of the concrete existence of objects in the external world. Yet here again closer scrutiny undermines commonsense impressions. Under a microscope the material of the rock is revealed to be a tangle of interlocking crystals. An electron microscope can uncover the individual atoms, spaced out in a regular array with large gaps in between. Probing into the atoms themselves, we find that they are almost entirely empty space. The tiny nucleus occupies a mere trillionth (10^{-12}) of the atom's volume. The rest is populated by a cloud of neither-here-nor-there ephemeral electrons, pinpricks of solidity whirling about in oceans of void. Even the nucleus, on closer inspection, turns out to be a pulsating package of evanescent particles. The apparently concrete matter of experience dissolves away into vibrating patterns of quantum energy.

There is no doubting the strong mystical element that underlies much of the new physics. The old view of the universe as a clockwork mechanism slavishly unfolding along a predetermined pathway, embedded in an absolute spacetime framework, has been swept away. In its place is a collection of images, each reflecting one aspect of commonsense experience, but failing to connect together in an orderly way. Is an electron a wave or a particle? Both forms conjure up a clear mental image, but we cannot relate to any one entity for which the answer is 'both'. Nor can we easily picture the idea of space being curved or undergoing expansion. Space we associate with emptiness, and warped emptiness is a mental obstacle few can surmount.

The mystical appeal of the new physics has endeared it to many people of a religious or philosophical turn of mind, who see in the discoveries now being made a release from the materialistic, impersonal world that modern technological society has created. Timewarps and quantum weirdness open up rich possibilities for the belief that there is more to the world than meets the eye. Especially attractive is the strong holistic flavour of the new physics. Much of the recent disillusionment with science stems from a reaction to traditional scientific reductionism where the world is analysed coldly into its simplest components.

The idea that everything can be understood by reducing it to its constituent parts has exercised a powerful influence on scientific thinking for several centuries. Newton realized that the complexities of motion could be understood by considering simple, small bodies acted upon with forces produced by other simple, small bodies. Though the behaviour of a falling leaf may seem bafflingly complicated, the motions of individual particles ought in principle to conform to simple mathematical principles.

The pinnacle of Newtonian reductionism came with a famous statement by Pierre Laplace:

> 'An intelligence knowing, at any given instant of time, all forces acting in nature, as well as the momentary positions of all things of which the universe consists, would be able to comprehend the motions of the largest bodies of the world and those of the smallest atoms in one single formula, provided it were sufficiently powerful to subject all data to analysis; to it, nothing would be uncertain, both future and past would be present before its eyes.'

Thus, if only one knew the exact positions and motions of all the particles of matter in the universe, then in principle the entire past and future behaviour of everything could be determined. The idea that all behaviour is laid down in this rigid way demolishes the notion of free will, and conjures up an image of sterile, mindless cosmos. It becomes still more offensive when living systems are concerned: the attempt to reduce all living things to nothing but moving mounds of atoms evolving according to blind chance has, more than anything, created the impression of science as a soulless and dehumanizing pursuit.

It is against this traditional reductionist background that the new physics stands out in such stark contrast. The quantum factor throws out Laplace's determinism, and it denies that the world can be understood in terms of its components alone. In the next chapter we shall see how two particles, even when apparently isolated by great separation, are nevertheless linked into a coherent pattern of behaviour. Quite generally, when any sort of measurement or observation is performed in quantum physics, the reality of the subatomic particle cannot be untangled from the environment it inhabits. We saw how, in Young's two-slit experiment, the behaviour of a particle as small as an electron depends on whether one or two slits are available to it as it passes through the screen. In some mysterious way the electron encodes information about a comparatively vast structure in its neighbourhood, and responds accordingly. Similarly, the direction in which a particle's spin points is inseparable from the measurement strategy adopted by the experimenter. Evidently the macroscopic and microscopic worlds are intimately interwoven. There is no hope of building a full understanding of matter from the constituent particles alone. Only the system *as a whole* gives concrete expression to microscopic reality. The big and the small co-exist. One does not subsume wholly the other, nor does the other wholly 'explain' the one.

One of the chief casualties of reductionist science was mind. In attempting to reduce all systems to the activities of simple components, some scientists came to believe that mind is nothing but the activity of the brain, which is nothing but a pattern of electrochemical impulses, which in turn is nothing but the motion of electrons and ions. This extreme materialistic philosophy reduces the world of human thoughts, feelings, and sensations to a façade.

The new physics, by contrast, restores mind to a central position in nature. The quantum theory, as it is usually interpreted, is meaningless without introducing an observer of some sort. The act of observation in quantum physics is not just an incidental feature, a means of accessing information already existing in the external world; the observer enters the subatomic reality in a fundamental way and the equations of quantum physics explicitly encode the act of observation in their description. An observation brings about a distinct transformation in the physical situation. When someone looks at an atom, the atom jumps in a characteristic fashion that no ordinary physical interaction can mimic. Common sense may have collapsed in the face of the new physics, but the universe that is being uncovered by these advances has found once more a place for man in the great scheme of things.

3
Reality and the Quantum

A maze of paradox

In the summer of 1982 an historic experiment took place at the University of Paris. French physicist Alain Aspect and his co-workers had set out to see if they could cheat the quantum. At stake was not only science's most successful theory, but the very foundation of what we believe to be 'reality'.

Like many crucial tests in physics, the Paris experiment was rooted in a paradox, one that has perplexed and intrigued scientists and philosophers alike for nearly fifty years. The issue concerns the central feature of quantum physics, uncertainty. The famous Heisenberg uncertainty principle seems to put paid to any simple, intuitive picture of the atomic world in which particles move along well-defined paths under the action of forces. A particle such as an electron will zig-zag about unpredictably, its motion impossible to follow, or even meaningfully describe, in detail.

Before the advent of the quantum theory, the physical universe was regarded as rather like a huge clockwork mechanism, its behaviour legislated in every detail by the unassailable logic of cause and effect embodied in Newton's laws of motion. Newton's laws are, of course, still good for most everyday phenomena. They direct the bullet to its target and supervise the planets in their orbits with geometrical precision. But we now know that, on an atomic scale, things are very different. The familiar, orderly behaviour of macroscopic bodies gives way to rebellion and chaos. What we used to regard as solid objects are found to be a ghostly mosaic of quivering energy. Quantum uncertainty ensures that you can't know everything about a particle all of the time. If you try to pin a particle down (metaphorically), it slips from your grasp.

This intangible quality of quantum particles has proved a source of great discomfort to the physicists involved in developing the theory. In the

1920s the new quantum mechanics seemed to be a maze of paradox. Though Werner Heisenberg and Erwin Schrödinger were the chief architects of the theory's technical structure, the interpretation of the theory fell to Max Born and, especially, to Niels Bohr. Bohr, a Danish physicist, had been the first person to fully recognize that the quantum theory applied to matter as well as radiation, and in later years he became the leading authority and spokesman for the physics community on the conceptual foundations of quantum mechanics. His Institute in Copenhagen became the focus for research in quantum physics for a decade. Bohr once remarked to his colleagues, 'If a man does not feel dizzy when he first learns about the quantum . . . he has not understood a word.' In his book *Physics and Philosophy*, Heisenberg recalled the early misgivings he had about the meaning of the new quantum mechanics:

> 'I remember discussions with Bohr which went through many hours till very late at night and ended almost in despair; and when at the end of the discussion I went alone for a walk in the neighbouring park, I repeated to myself again and again the question: Can nature possibly be so absurd as it seemed to us in these atomic experiments?'

The most notable malcontent was Einstein. Although he had a hand in formulating the quantum theory, Einstein never felt happy about it, believing it to be either wrong, or at best a half-truth. 'God does not play dice' is one of his most famous utterances. He maintained that behind the quantum world of unpredictable fuzziness and disorder lay a familiar classical world of concrete reality in which objects really possess well-defined properties such as location and speed and move according to deterministic laws of cause and effect. The madhouse of the atomic domain, Einstein declared, is not fundamental. It is a façade. At a deeper level of description, sanity reigns.

Einstein tried to find this deeper level in a running debate with Bohr, the strongest articulator of the 'party line' which maintains that quantum fuzziness is inherent in nature, and irreducible. Einstein sallied forth with many an imaginative assault on quantum uncertainty, trying to invent hypothetical experiments – 'thought experiments' as they are known – designed to expose a logical flaw in the official view. Again and again Bohr would counter-attack and shoot Einstein's argument down.

On one memorable occasion, at a conference where most of Europe's leading physicists had gathered to hear about the latest advances in the, then novel, quantum theory, Einstein directed his attack towards a variant of the uncertainty principle dealing with how precisely the energy of a particle can be determined along with the instant of time at which it possesses that energy. He had devised a particularly ingenious scheme to circumvent the energy–time uncertainty. The basis of his idea was to measure energy accurately using its weight. Einstein's famous $E = mc^2$ relation assigns a mass m to a quantity of energy E, and mass can be measured by weighing it.

This time Bohr was visibly rattled, and witnesses tell of him accompanying Einstein back to the conference hotel in agitated mood. After a sleepless night in which Bohr analysed Einstein's argument in detail, he addressed the reassembled conference the following day, triumphant. In deploying his argument against quantum uncertainty Einstein had, ironically, overlooked an aspect of his own theory of relativity. This theory requires that time is warped by gravity. As gravity must be present in order to carry out a measurement of weight, the warping effect cannot be ignored. Bohr demonstrated that when this effect is properly taken into account, uncertainty returns at the same level as always.

The Einstein–Podolsky–Rosen experiment

The most enduring of Einstein's thought experiments did not surface until 1935, when, together with his colleagues Boris Podolsky and Nathan Rosen, he published a seminal paper in *The Physical Review* which to this day remains the most cogent formulation of the paradoxical nature of quantum physics. What the Einstein–Podolsky–Rosen experiment addressed was, in essence, the problem of whether a particle can have both a position and a momentum at the same time. The challenge facing Einstein and his colleagues was to devise a scheme in which it appears that both these quantities can, in principle at least, be measured to any desired degree of accuracy.

It had by that time come to be accepted that any direct attempt to determine both the position and the momentum of a particle was doomed to failure. The reason is simple: when you go to measure the position, the very effect of the measurement throws out the momentum in an indeterminable way. And a measurement of momentum destroys any previous information about position. Each type of measurement is incompatible with the other and blurs the result of the other. If Einstein was to succeed in finding a way to determine both properties simultaneously it had to involve a more subtle strategy.

Boiled down to its essentials, what Einstein, Podolsky, and Rosen came up with was this. Given that you can't *directly* determine both the position and momentum of a particle at the same time, what you need is a second, accomplice, particle. With *two* particles, more quantities can be measured at once. If the motion of the two particles can be linked in advance in some way, then measurements performed simultaneously on both particles might enable the experimenter to sneak a look beneath the veil of quantum uncertainty that Bohr insisted could never be lifted.

The principle involved is familiar enough. In a game of snooker or pool, when the cue ball strikes another, the two balls move off in different directions. Their motion is not, however, random, but precisely fixed by the law of action and reaction. If you measure the momentum of one ball you can deduce that of the other (which might by then be far away) without ever directly observing the other ball. Now the law of action and reaction also applies to quantum particles. All that one basically needs to do, then, is to let two quantum particles, 1 and 2, come together, interact, and separate to a great distance. At this point the momentum of particle 1 can be measured. From the law of action and reaction this enables the momentum of particle 2 – the particle of interest – to be accurately deduced. The measurement, to be sure, will have rendered the position of particle 1 uncertain, but that is of no concern. It cannot have affected the position of particle 2 – the one we want to get at – because particle 2 is far away; in principle it could be light years away. If, simultaneously, one also measures the position of particle 2 directly, then both the position and the momentum of particle 2 are known at the same instant. We will have succeeded in out-manoeuvering the uncertainty principle!

The Einstein–Podolsky–Rosen argument is founded on two crucial assumptions. The first is that a measurement performed in one place cannot instantaneously affect a particle in another, very distant place. The basis for this belief is, first, that interactions between systems tend to decline with distance. It is hard to imagine two electrons several metres, let alone light-years apart influencing each other's position and motion in an intricate way. Einstein dismissed such an idea, referring to it as 'ghostly action at a distance'.

An important reason for this rejection was Einstein's belief that no signal or influence can travel faster than light. This is a key result of the theory of relativity and is not to be relinquished lightly. Among other things, the absence of faster-than-light signalling is a vital element in establishing a common definition of past and future throughout the universe. Breaking the light barrier is tantamount to signalling backwards in time, a prospect replete with paradox.

The second fundamental assumption made by Einstein and his colleagues was that of 'objective reality'. They assumed that a quality such as the position or momentum of a particle exists objectively, even if the particle is at a distant place and the quality concerned is not directly observed. This is where Einstein's outlook departed from that of Bohr. According to Bohr, you simply cannot ascribe attributes like position and momentum to a particle unless you actually carry out an observation of the particle. Measurement by proxy will not do. Using an accomplice particle is cheating.

At this stage both Bohr and Einstein could maintain their entrenched positions. What was needed was a variant of the thought experiment that would enable a direct test to be made on whether or not the uncertainty principle was being violated in a practical experiment. In the 1960s, John Bell of the Centre for European Nuclear Research (CERN) discovered a way to do this. He took the two basic assumptions of Einstein, Podolsky, and Rosen – no faster-than-light signalling and the existence of objective reality – and used them to deduce the most general relationships between measurements on particle 1 and measurements on particle 2; not just position and momentum, but other properties such as direction of spin. He found that certain sorts of measurements could be performed which would discriminate between Einstein's and Bohr's positions. That is, the two above assumptions could lead to certain experimental predictions which would not be met if quantum mechancis *à la* Bohr, with its inherent uncertainty, were correct. If, then, a real experiment were performed, a direct test of quantum uncertainty could be carried out.

Bell encoded the essential difference between the two rival theories in the form of a mathematical statement known, appropriately enough, as Bell's inequality. Put simply, if Einstein were right, Bell's inequality would hold in the actual experimental results. If Bohr were right, the inequality would be violated. The ball was now in the experimenter's court.

The collapse of naive reality

A practical test of Bell's inequality could not be carried out in the 1960s. The main problem was the precision of the available technology. To be certain that two separated particles are not communicating in the conventional way, it is necessary to carry out operations on both particles within such a short interval of time that there could not be long enough for any signals travelling at the speed of light (or less) to pass between them. For particles several feet apart, this means the operations concerned must take no longer than a few thousand-millionths of a second.

During the 1970s a number of experimental groups set up two-particle experiments of various sorts, though none achieved the accuracy required to make the results watertight. Finally, Alain Aspect in Paris introduced a number of refinements, and in 1981 began a series of experiments in which the polarization angles of two oppositely moving photons emitted by a single atom were simultaneously examined. The programme culminated in an experiment performed in the summer of 1982 which for the first time really seemed to clinch it. The results were unequivocal. Einstein was wrong. Quantum uncertainty cannot be bypassed. It is intrinsic and irreducible. Naive reality – the reality of particles really possessing well-defined qualities in the absence of observation – cannot be sustained. Aspect had put the last nail in the coffin of commonsense physics.

The way in which the experiment exposes the difference between the quantum theory and any alternative 'realistic' theory is of some interest. The experimenters set out to see to what extent the results of measurements on one photon were correlated with those of the other. According to Bell's inequality, theories of the 'realistic' type predict a certain maximum level of correlation. Quantum mechanics, by contrast, predict a greater degree of correlation, as though the two particles are co-operating by telepathy in some way to an unnatural extent. The results showed a correlation in excess of the maximum permitted by Bell's inequality and therefore confirmed the intrinsic uncertainty in quantum physics.

The situation can be compared to two individuals sitting back to back, tossing coins simultaneously. If the tosses are fully random, there will be no correlations expected between the two coins. The occurrence of heads on one coin is equally likely to occur with either heads or tails on the other coin. Suppose, however, that the tosses were not purely random, so that a heads result for one coin was more likely than not to concur with a heads on the other, and similarly for tails. Observations would show a definite positive correlation between the results of the two coins. In the two-particle experiments the activities of the particles are not independently random because both particles possess a common origin. Some correlation is therefore expected. The precise degree of correlation provides the crucial test.

At first sight it might seem as if the Aspect experiment provides a means of achieving faster-than-light signalling. In the language of coin tossing, if my heads is likely to occur with your heads, it seems I can send you a message, even though you cannot see my coin, by arranging a simple code, e.g. heads for dot, tails for dash. If the correlation is less than 100 per cent the message would be 'noisy', but could eventually be transmitted accurately with enough repetition.

Further thought, however, shows this possibility to be illusory. The outcome of each of my coin tosses, although correlated with your coin tosses, is still nevertheless completely unpredictable since I cannot in advance make my coin come down heads or tails to order. *If* it comes down heads I know that yours probably shows heads too, but that is no use to me. I have no control over the sequence of dots and dashes being sent; the message degenerates into white noise.

The weird nature of quantum reality

Several months after Aspect published the results of his experiment I had the privilege of making a BBC radio documentary programme about the conceptual paradoxes of quantum physics. The contributors included Aspect himself, John Bell, David Bohm, John Wheeler, John Taylor, and Sir Rudolph Peierls. I asked all of them what they made of Aspect's results and whether they thought that commonsense reality was now dead. The variety of answers was astonishing.

One or two of the contributors felt no surprise. Their faith in the official view of the quantum theory as enunciated long ago by Bohr was so strong that they felt the Aspect experiment merely provided confirmation (albeit welcome confirmation) of what was never seriously in doubt. On the other hand, some were not prepared to leave it at that. Their belief in commonsense reality – the objective reality sought by Einstein – remained unshaken. What would have to go, they argued, was the assumption that signals could not travel faster than light. There must be some 'ghostly action at a distance' after all. Bohm already has a ready-made theory that incorporates such 'non-local' effects.

And what of the time-signalling paradoxes? Well, perhaps something prevents such signals being sent in a controlled way? The issue was left vague.

Although not all physicists seem to accept the overthrow of naive reality, Bohr's position remains the official view, and has undoubtedly been strengthened by Aspect's results. If this position is adopted it has some very profound implications for the nature of the physical world.

First, the two-particle arrangement described in the foregoing reveals that the reality of a particle 'over there' is indissolubly linked with the reality of a particle 'over here'. The simplistic assumption that just because two particles have moved a long way apart we can consider them as separate and independent physical entities is badly wrong. Unless separate measurements have taken place on both the particles, they remain part of a unified whole. What we mean by reality is defined only by the total experimental arrangement, which could be spread out over a large region. Furthermore, although in the Aspect experiment such a two-particle 'holistic' system is deliberately set up in a controlled way, all the time particles are continually interacting and separating as a result of their natural activity. The non-local aspect of quantum systems is therefore a general property of nature, and not just a freak situation manufactured in the laboratory.

Some people have emphasised that quantum physics implies a world in which individual particles of matter do not really exist in their own right as primary entities. Instead, only the collection of all particles treated as a whole, including those that go to make up the measuring apparatus, has the status of 'reality'.

The more traditional view of reality based on classical Newtonian physics is quite different. According to Newtonian philosophy, matter is made up of particles, but the particles are regarded simply as building blocks that can assemble into larger units. This picture is appealing because we can easily visualize myriads of these elementary particles like solid balls, locked together to form a familiar object such as a rock. All the properties possessed of the rock can then be attributed to the atoms, or whatever basic building block is fashionable. The rock is *made up of* elementary particles and the particles are simply fragments of the rock. Nothing more. The German physicist Otto Frisch, the discoverer of nuclear fission, describes the classical picture as follows:

> 'It takes the line that there is definitely an outside world consisting of particles which have location, size, hardness and so on. It is a little more doubtful whether they have colour and smell; still, they are *bona fide* particles which exist there whether or not we observe them.'

We might call this classical philosophy 'naive realism'.

In quantum physics this simplistic classical relationship between the whole and its parts is totally inadequate. The quantum factor forces us to perceive particles only in relation to the whole. In this respect it is wrong to regard the elementary particles of matter as *things* that collectively assemble to form bigger things. Instead, the world is more accurately described as a network of *relations*.

To the naive realist the universe is a collection of objects. To the quantum physicist it is an inseparable web of vibrating energy patterns in which no one component has reality independently of the entirety; and included in the entirety is the observer.

The American physicist H. P. Stapp has expressed the quantum concept of particle in these words:

> 'An elementary particle is not an independently existing unanalysable entity. It is, in essence, a set of relationships that reach outward to other things.'

One is reminded of the words of William Blake: 'To see a world in a grain of sand . . .' We must envisage all matter and energy everywhere encompassed in a unified existence.

A further consequence of quantum physics concerns the role of the observer, the person who actually carries out the measurements. The fuzziness of quantum uncertainty does not carry through to the actual observations we make, and so at some stage in the chain between the quantum system of interest, up through the experimental equipment, the dials and meters, our own sense organs and brains, and finally our consciousness, something must happen to dispel the fuzziness. The rules of quantum physics are quite definite on this point. In the absence of an observation a quantum system will evolve in a certain way. When an observation is made, an entirely different type of change occurs. Just what produces this different behaviour is not clear, but at least some physicists insist that it is explicitly caused by the mind itself.

On this mystical note we leave the problems and paradoxes of quantum physics. Whatever the controversies that remain over its conceptual foundations, there is no lack of agreement that, in its applications, the quantum theory works brilliantly. In particular, it forms the basis for all our description of the world of particle physics, the world in which the superforce lies buried.

4
Symmetry and Beauty

'Beauty is truth, truth is beauty.'
John Keats

Mathematics as the language of nature

When I lecture to freshers on 'The Concepts of Modern Physics' I always tell them that physics is beautiful because it is described by such simple mathematical laws. The remark usually produces howls of derision. The reason is, of course, that to a student struggling with undergraduate mathematics courses, the equations of physics seem horribly complicated and opaque. What they have yet to appreciate is that mathematics is, among other things, a language. When that language has been learned, immensely complicated things can be elegantly summarized in the mathematical equivalent of a one-liner.

In this respect, mathematics differs little from other technical languages (though it is immeasurably more powerful and comprehensive). Imagine, for example, trying to explain an investment scheme to somebody in ordinary English, without being able to use the words capital, interest, or inflation. Or envisage describing the workings of a car engine without ever mentioning pistons, cam-shafts, gaskets, or carburettors.

Anyone who has ever heard two mathematicians in conversation might well deduce that they are talking in code, and in a sense they are. Like all codes, once you know the key, complex information instantly becomes simple. If a coded message is intercepted, it will be recognized as an orderly arrangement of information, but the actual information content is hidden in a meaningless jumble of cyphers. A mathematical formula is rather like a code, with an input and an output. Take, for example, the formula n^2, where n is any natural number 1, 2, 3, 4, . . . Inserting the values for n in sequence yields 1, 4, 9, 16, . . . In this case the code is not very hard to crack, and most people could work backwards from the answer 1, 4, 9, 16, . . . to deduce the formula n^2 and the input 1, 2, 3, 4, . . .

But make the formula just a bit more complicated, and cracking the code rapidly becomes a daunting task. The reader is invited to fit a formula to the sequence 2, 4, 6, 9, 12, 17, 20, 25, 28, 31, 34, . . .

Perhaps the greatest scientific discovery of all time is that nature is written in mathematical code. We do not know the reason for this, but it is the single most important fact that enables us to understand, control, and predict the outcome of physical processes. Once we have cracked the code for some particular physical system, we can read nature like a book.

The realization that at a fundamental level the laws of nature are written in mathematical code dawned only slowly on mankind. Early astrologers deduced simple numerical relationships that govern the motion of the sun, moon, and stars, and help in predicting eclipses. Pythagoras discovered that the musical tone of a plucked string bears a precise numerical relationship to its length. It was only in the Middle Ages, however, that the first systematic attempts were made to unravel nature's mathematical code. In the fourteenth century Oxford scholars deduced the interesting fact that the vertical distance travelled by a body dropped from rest is proportional to the square of the elapsed time, t^2. But the general acceptance of this fact had to await the seventeenth century and the work of Galileo and Newton. Other, related facts were that the period of a swinging pendulum is independent of the amplitude of its swing, but proportional to the square root of its length, and that a projectile always follows a precise geometrical curve known as a parabola. Kepler deduced mathematical relationships governing the motions of the planets, such as the fact that the squares of the orbital periods are proportional to the cubes of their mean distances from the sun.

All this culminated in Newton's establishment of the laws of mechanics and gravity. He found that gravity complies with a particularly simple mathematical formula known as the inverse square law. This law relates the force of gravity to the distance r from the centre of a spherical body by the relationship $1/r^2$. Later experiments with electric and magnetic forces revealed that they, too, obey inverse square laws.

In the eighteenth and nineteenth centuries the mathematical foundation of physics expanded enormously. In many cases new mathematical topics were invented to cope with the burgeoning demands of the physicists. This century the mathematical development of physics has gone much further, and incorporated many abstract branches of mathematics such as non-Euclidean geometry, infinite-dimensional vector spaces, and group theory.

Things which may appear very complicated or unsystematic on the surface can turn out to be expressions of very simple mathematical relationships once the code is broken. When the physicist explores the world of nature, he may encounter what at first seems to be very involved, or even random behaviour. Later, with the help of a proper mathematical theory, that behaviour can turn out to be a manifestation of disarmingly simple mathematics.

No better historical example can be given than the case of planetary motions in the solar system. That the planets move in the sky in a relatively ordered way is familiar to anyone with even a casual interest in astronomy. Detailed study, however, reveals that individual planets differ greatly in their motion. Mars, for example, usually moves from east to west across the fixed backdrop of distant stars, but will on occasions turn around and move temporarily eastwards. 'However, the outer planets move much more slowly than the inner ones.' Closer analysis reveals many more complicated features.

A once favourite model of the solar system was the one devised by Claudius Ptolemy in the second century A.D., based on the assumption that the Earth is fixed at the centre of the universe and the planets are attached to rigid concentric spheres which rotate at various speeds. As more accurate observations demanded ever more intricate motions, it became necessary in this system to augment the original spheres with other smaller spheres turning within them, so that the combined motions of two or more rotations could reproduce the observed planetary activity. By the time that Copernicus discovered the true organization of the solar system in the sixteenth century, the Ptolemaic system of spheres had become immensely contrived and complicated.

The scientific revolution which accompanied the work of Galileo and Newton provides a classic example of how what seems to be a complicated mess can, using a more sophisticated mathematical model, be revealed as elegant simplicity. Newton's main step was to treat the planets as material bodies moving through space, subject to the physical laws of motion and gravity – laws he had himself already deduced. As a result he was able to describe both the sizes and shapes of the planetary orbits, as well as their periods. All was in nearly perfect agreement with observation. The point is that Newton's laws of gravity and motion are both very simple mathematically, even by high school standards. Nevertheless, when combined they lead to a rich and complex variety of motions.

The example of planetary motion also serves to illustrate an important point about the physical world. I am often asked why it is, if the laws of physics are so simple, that the universe is so complicated. The answer comes from a proper understanding of what is meant by a law. When a physicist talks about a law he means some sort of restriction in the behaviour of a class of systems. For example, a simple law states that all pitched baseballs follow parabolic paths. The law can be checked by examining many different cases of pitched baseballs. The law does not say that all the paths are the same. If they were the game would be dull indeed. Some parabolas are low and flat, others high and curved, and so on. While all the paths belong to the class of curves called parabolas, there is an infinite variety of parabolic curves from which to choose.

What, then, determines the particular parabolic path that any given baseball will follow? Here is where the pitcher's skill comes in, because the shape of the path will depend on the speed of delivery and the angle of pitch. These two additional parameters, known as 'initial conditions', must be specified before the path is uniquely selected.

A law would be useless if it was so restrictive that it permitted only one possible pattern of behaviour. It would then become a *description* of the world rather than a true law. In the real world the richness and complexity of activity can still be based on simple laws because there is an unending variety of initial conditions to go with them. In the solar system the laws require that all planetary orbits be elliptical, but the precise size and elongation of each ellipse cannot be deduced from the laws. They are determined by the initial conditions, which we do not know in this case because they depend on how the solar system formed in the first place. The same laws also describe the hyperbolic paths of comets, and even the complicated trajectories of spacecraft. Thus, the simple mathematical principles discovered by Newton can support a vast range of complex activity.

Beauty as a guide to truth

Beauty is a nebulous concept, yet there is no doubt that it provides a source of inspiration for professional scientists. In some cases, when the road ahead may be unclear, mathematical beauty and elegance guide the way. It is something the physicist feels intuitively, a sort of irrational faith that nature prefers the beautiful to the ugly. So far this belief has been a reliable and powerful travelling companion, in spite of its subjective quality.

Heisenberg once made the following remark to Einstein:

'If nature leads us to mathematical forms of great simplicity and beauty . . . that no one has previously encountered, we cannot help thinking that they are "true", that they reveal a genuine feature of nature.'

He went on to discuss 'the almost frightening simplicity and wholeness of the relationships which nature suddenly spreads out before us', a theme echoed by many of his contemporaries. Paul Dirac went so far as to declare that 'It is more important to have beauty in one's equations than to have them fit experiment.' The point that Dirac was making is that a leap of creative imagination can produce a theory which is so compelling in its elegance that physicists may become convinced of its truth before it is subjected to experimental test, and even in the face of what appears to be contradictory experimental evidence.

In his delightful book *Dismantling the Universe*, science writer Richard Morris picks up this point:

'There are more parallels between science and the arts than immediately strike the eye. Like artists, scientists often have unique styles. Furthermore, their ideas of what a good scientific theory should be like are strangely reminiscent of artistic convictions. . . . A correct theory is one that can presumably be verified by experiment. And yet, in some cases, scientific intuition can be so accurate that a theory is convincing even before the relevant experiments are performed. Einstein – and many other physicists as well – remained convinced of the truth of special relativity even when . . . experiments seemed to contradict it.'

Morris relates a story of how Einstein reacted to the news that a crucial prediction of his general theory of relativity had been confirmed in an astronomical experiment. Einstein seemed unmoved. Asked how he would have reacted had the experiment contradicted the theory, he replied, 'Then I would have been sorry for the dear Lord – the theory is correct.'

Mathematical elegance is difficult to communicate to those without some mathematical background, but I shall try. Look at the curve in Figure 6. Though the shape is basically smooth and relatively featureless, it does not immediately correspond to anything encountered in daily life. If you were asked to remember such a curve and reproduce it *exactly* on some later occasion, the task would be hopeless. You could do this with, say, a circle, or even a more complicated shape that you recognized, such as an ellipse (which is a circle viewed obliquely), but the curve shown in Figure 6 has much more structure than a circle. Both the slope and the curvature of the curve change along its length in a systematic way, which is nevertheless hard to define exactly.

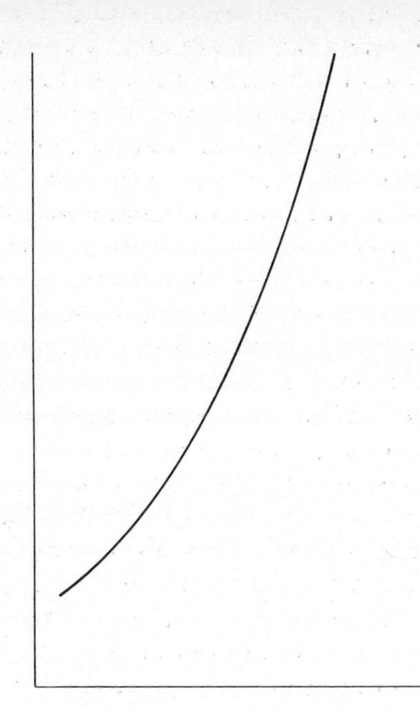

Figure 6. The exponential curve. The shape of this curve encodes some deep mathematical properties that recur in a wide range of physical situations. Regarded as a graph, the exponential could represent, for example, the unrestrained growth of human population.

A mathematician, however, will recognize this curve and know how to encode its entire range of properties so that he can easily remember it and reproduce it to any desired degree of accuracy on a future occasion. The curve is, in fact, a graph of the so-called 'exponential' function, denoted e^x, which reappears again and again in a wide range of scientific problems. The mathematician will know that this function can be derived from the formula $(1 + x/n)^n$ in the limit that n becomes infinitely large, and so armed with a calculator he can compute each point on the graph to any required precision.

'The exponential function is one of the most elegant relationships known to man,' claims the mathematician. Why?

Suppose we consider the gradient of the curve at each point. It starts off shallow on the left and grows steeper and steeper. Plot a graph, not of the exponential function, but of the gradient of the exponential function. What shape is this? It turns out to be identical to the exponential function itself. An exponential curve is one that is everywhere equal to (or at least proportional to) its own slope. This makes the exponential function of great importance in describing simple forms of growth, such as that of unrestricted breeding populations, in which the gradient, which is a measure of the rate of growth, is proportional to the population itself. This is approximately true of the human population in some parts of the world today.

Further hidden elegance can be found in the exponential curve. Look at the shape shown in Figure 7. This time the curve does suggest something familiar: a wave. The curve is known technically as a sine wave, written $\sin x$, and it can also be reproduced algebraically by computing from a formula.

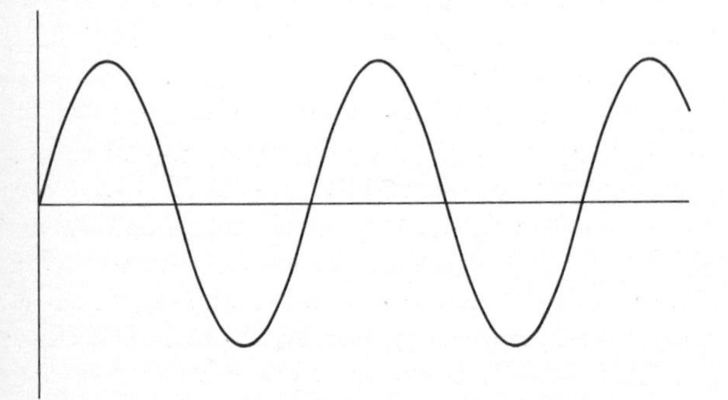

Figure 7. The sine curve. This characteristic shape possesses mathematical properties closely related to the exponential curve shown in Figure 6. It describes a wide range of physical phenomena, including wave motion and periodic oscillations.

Superficially the sine wave bears very little resemblance to the exponential. The sine wave is periodic – it goes up and down regularly – whereas the exponential curve always rises faster and faster. The connection between the two curves is revealed by plotting the gradient of the sine curve. The result is another sine curve, but displaced by one-quarter of one wavelength to the right. This is called a cosine curve. If the gradient of the cosine curve is plotted, we shift another half-wavelength to the right, giving a curve which is the same as the sine curve drawn upside down. After two more gradient operations we finally get back to the original curve. The exponential and sine (or cosine) curves thus share an important symmetry property which connects the shape of the curve to the shape of the gradient of the curve.

This deep connection between e^x and $\sin x$ is made fully explicit in the theory of complex numbers, in which the usual number system is extended to include the square roots of negative numbers. One finds that when x is the square root of a negative number, e^x becomes a mixture of sine and cosine waves. It is then no surprise to find that physical systems which display exponential behaviour are also likely to display periodic 'sinusoidal' behaviour. One example is the so-called harmonic oscillator. This could be a pendulum or simply a mass attached to a spring. If the mass is displaced slightly, it will oscillate back and forth, driven by the periodic force of the spring. The position of the mass will vary in a way that is described by the sine function shown in Figure 7. The reason for this can be traced to the law of force for springs, i.e. that the force of the spring on the mass is directly proportional to the displacement of the mass from the equilibrium position. The direction of the force is such as to try to pull the mass towards the equilibrium position. Thus, if the spring is extended it is a pulling force; if it is compressed it is a pushing force.

Suppose now that the force law were the same, but the force acted in a direction away from rather than towards the equilibrium position. Very different behaviour would result. The mass would follow an exponential curve, accelerating away faster and faster in the same direction. Springs do not do this of course, but other systems do. Sometimes a system will oscillate like a sine wave under certain conditions, and then shoot off exponentially under other conditions.

Finding hidden relationships and symmetries like this through mathematical analysis forms part of the intellectual skills of all professional physicists. Often it is only by a revolutionary change in the mathematical formulation of the description that the more subtle symmetries are exposed. This happened in the passage from Ptolemaic cosmology to Newtonian mechanics, and it also happened much later to Newtonian mechanics itself.

The laws of Newton were completely reformulated by the French physicist Joseph-Louis Lagrange and the Irish physicist William Rowan Hamilton in the nineteenth century. In both cases the mathematical description was adjusted to take advantage of the simplicity and elegance embodied in Newton's mechanics. Hamilton's work in particular contained an unexpected pointer to the quantum revolution that was to sweep the whole of classical physics completely away. But that was yet to come.

The central problem of mechanics is to understand, describe, and predict the trajectories or paths followed by particles of matter under the action of applied forces. These paths will obviously have an infinite variety of shapes depending on the nature of the forces that are acting. An apparently completely different problem concerns the paths followed through a transparent medium by light rays. Light is not subject to Newton's laws of mechanics, though it is a familiar fact that light paths can be bent by passing them through material of varying density. For example, a stick looks bent when placed in a pond. The reason for this is that light waves are slowed up in dense materials so that, roughly speaking, the wavelets at different points on a wavefront get out of step with each other as they encounter media of different densities. The end result is that, in most cases, a given light ray will follow a path that minimizes the time it takes to get from place to place. Thus, the shape of a light ray can be understood in terms of the theory of waves which travel at varying velocities according to the nature of the medium through which they are passing.

When Hamilton had reformulated Newton's mechanics, he found that the most succinct expression of the laws of motion were contained in a mathematical statement identical to that of the minimum time principle for light waves. Roughly, particles try to get from place to place along the easiest (which in most cases means the fastest) route available. It was therefore found that both material particles and light waves, despite their totally different character and behaviour, actually move in a way which is more or less identical in a certain mathematical sense.

This astonishing result, which comes entirely from the attempt to rewrite the laws of mechanics in a new mathematical form, reveals a deep harmony in nature that is suggestive of some hidden principles at work. With the benefit of hindsight we can now see what these principles are. The close relationship between particle motion and the motion of light waves suggests that there is, perhaps, a wavelike quality associated with material particles. These 'particle waves' were discussed in Chapters 2 and 3, and form the starting point of the quantum theory. Thus Hamilton's mathematical juggles can be seen as pointing the way forward to a new wave theory of matter.

Symmetry

The concept of symmetry is a familiar and important one in daily life. Many human products are deliberately built to be symmetric, for either aesthetic or practical reasons. A ball is symmetric because it looks the same if you rotate it about its centre in any direction. A chimney also remains unchanged under a more restricted set of rotations, namely those around a vertical axis running down the centre of the chimney.

In nature, too, symmetry abounds. A snowflake displays a most remarkable hexagonal symmetry. Crystals display characteristic geometrical shapes, e.g. the cubical form of salt crystals, which reflect the underlying regularity of the atomic structure. The falling raindrop is a perfect sphere, and if frozen can form into the spherical ice ball we call a hailstone.

Another symmetry frequently encountered in nature and in human design is reflection symmetry. The human body is approximately reflection-symmetric about a vertical centre line. In a mirror, right and left hands, and other features, are interchanged, but the image we see is still recognizable. Many architectural designs are reflection-symmetric, such as archways and cathedrals.

There is an intimate connection between geometrical symmetry and what are known in physics as conservation laws. Conservation laws tell you that certain things remain unchanged with time. In American football, the number of players on the field is conserved. Players may come and leave, but the total number remains fixed. In physics there is a law that in an isolated system energy must be conserved, as must momentum and angular momentum. These restrictions do not mean that the system cannot change, only that any change is always restricted so that these three quantities remain fixed. In the game of snooker, where the balls are approximately mechanically isolated on account of the smooth texture of the snooker table, the laws of energy and momentum conservation determine the directions and speeds of the balls.

The conservation laws follow directly from Newton's laws of motion, but the later reformulation of these laws by Lagrange and Hamilton bring out their significance much more suggestively. What they reveal is a deep and powerful connection between the conservation of a quantity and the presence of symmetry in the system of interest. For example, if the system is symmetric when rotated, then it follows from Hamilton's or Lagrange's equations that angular momentum will be conserved. A good illustration is provided by the sun's gravity. If the spherical sun were rotated about its centre, the Earth would continue unaffected along its orbit. The sun's gravitational field is symmetric in orientation, and so a simple rotation leaves it unchanged. This geometrical symmetry is translated into the physical result that the angular momentum of an orbiting planet is always constant. (This fact was actually discovered by Kepler in the seventeenth century, but he did not appreciate its true meaning.) Similar ideas apply to momentum and energy.

Symmetries such as rotations and reflections are easy to visualize and pleasing to the eye, but they do not exhaust the stock of symmetries employed by nature. Sometimes new and unexpected symmetries are discovered by physicists while exploring the mathematical description of a physical system. The symmetries are encoded in the mathematics in a cryptic and subtle way, and would not be readily apparent to someone observing the physical system itself. By manipulating the symbols in their equations, physicists can try to discover the full range of symmetries, including those that cannot be visualized.

A classic example occurred at about the turn of the century, and concerned the laws of motion of the electromagnetic field. Decades before, the work of Michael Faraday and others had established that electricity and magnetism were closely related, each feeding off the other. The effects of electric and magnetic forces could best be understood in terms of a *field*, a sort of invisible halo of influence emanating from matter and extending through space, able to act on electrically charged particles, electric currents, or magnets. You can feel this field when you bring two magnets close together and sense them pushing or pulling each other, even though they do not touch physically.

Later, in the 1850s, James Clerk Maxwell built upon this knowledge and wove together the electric and magnetic fields into a system of equations. Maxwell found initially that the equations looked unbalanced; the electric and magnetic parts did not come in quite symmetrically. He therefore added an extra term to make the equations look more pleasing and symmetric. The extra term could be interpreted as an effect that had been overlooked – the creation of magnetism by a varying electric field – but which turned out to actually exist. Nature obviously agreed with Maxwell's aesthetic sense!

Maxwell's introduction of the extra term led to the most profound consequences. First, it truly unified electricity and magnetism into a single field, the electromagnetic field. Maxwell's equations were the first unified field theory and the first step on the long road to the superforce. They demonstrated that what at first sight could appear as two distinct forces of nature are in fact just two different facets of a single, unified force.

Secondly, among the solutions to Maxwell's equations were some unexpected and exciting possibilities. The equations, it turned out, could be satisfied by various sine functions (symmetry again), which, as described earlier in this chapter, represent waves or periodic undulations. These *electromagnetic waves*, Maxwell concluded, would travel unaided through the field, and hence propagate through what appears to be empty space. His equations gave a formula for the speed of the waves in terms of electrical and magnetic quantities. Putting in the numbers, the speed worked out at about 300 000 km s^{-1}, which is the speed of light. The conclusion was inescapable: light must be an electromagnetic wave. It can indeed travel through free space, for that is how we see the sun.

Maxwell went further than this and predicted the existence of electromagnetic waves with other wavelengths too, and several years later his prediction was confirmed when Heinrich Hertz produced the first radio waves in the laboratory. Today we know that gamma rays, X-rays, infra-red, ultraviolet and microwaves are also electromagnetic waves. Maxwell's little bit of symmetry went a long way.

The discovery of electromagnetic waves, with their far-reaching ramifications for radio and, ultimately, the electronic revolution, is a splendid example, not only of the power of mathematics in describing the world and extending our knowledge of it, but also of the use of symmetry and beauty as a guiding principle. But the full implications of the symmetry of Maxwell's equations took fifty years to be appreciated.

Around the turn of the century Henri Poincaré and Heinrich Lorentz investigated the mathematical structure of Maxwell's equations, with particular regard to the symmetries that lay hidden among the symbols, symmetries which were then unknown. The celebrated 'extra term' that Maxwell had introduced to balance up his equations turned out to give the electromagnetic field a powerful but subtle form of symmetry that emerged only after careful mathematical analysis. It seems that only Einstein, with his superhuman insight, ever suspected such a symmetry on physical grounds.

The Lorentz–Poincaré symmetry is similar in spirit to the idea of geometrical symmetries such as rotations and reflections, but it differs in one crucial respect: nobody else had thought of mixing space and time together in a physical way. Space is space and time is time. The meeting of the twain in the Lorentz–Poincaré symmetry was odd and unexpected.

In essence, the new symmetry is rather like a rotation, but not in space alone. Instead, the rotation involves time. If the three dimensions of space are added to the one dimension of time to make a four-dimensional *spacetime*, then the Lorentz–Poincaré symmetry is a sort of rotation in spacetime. The effect of rotating in spacetime is to project some spatial length into time and vice versa. That Maxwell's equations are symmetric under such a peculiar space–time linking operation is deeply suggestive.

It took Einstein's genius to drive home the full implications. Space and time are not independent entities, they are interwoven. The subtle 'rotations' of Lorentz and Poincaré are not merely abstract mathematics, but *can occur* in the real world, through *motion*. The key to the weird spacetime 'projections' or distortions lies with the speed of light and electromagnetic waves in general, which also came out of Maxwell's equations. There is thus a deep relationship between the electromagnetic wave motion and the structure of space and time. When an observer moves at close to the speed of light, space and time become severely distorted, in a symmetric way, as described by the mathematical operations of Lorentz and Poincaré. This was the peculiar effect, so contrary to common sense, described in Chapter 2. It was from this major new insight into a subtle and hitherto unrecognized symmetry of nature that Einstein's theory of relativity was born, and with it the birth of the new physics which rocked the scientific community and changed the face of the twentieth century.

More abstract symmetries

The lesson of Lorentz and Poincaré is that formidable advances can be made in physics through mathematical exploration, particularly when the concept of symmetry is exploited. Even though the mathematical symmetries may be hard, or even impossible, to visualize physically, nevertheless they can point the way to powerful new principles of nature. Searching for undiscovered symmetries has become a major tool in helping the modern physicist advance his understanding of the world. We shall see how the superforce has emerged as the supreme example of symmetry in nature.

So far, the symmetries that have been discussed all involve space, or spacetime. But the concept of symmetry can be enlarged to include more abstract notions. As already explained, there is a close connection between symmetry and the conservation laws. One of the best-established conservation laws is that of electric charge. Charge can be both positive and negative, and the law of charge conservation says that the total quantity of positive charge minus the total quantity of negative charge cannot change. If a quantity of positive charge meets an equal quantity of negative charge, the two charges can neutralize each other to give zero net charge. Similarly, positive charge can be created so long as there is an equal quantity of negative charge created to go with it. But the appearance or disappearance of a net quantity of charge is strictly forbidden.

If electric charge is conserved, the question naturally arises as to the nature of the symmetry which is associated with it. One would search in vain to find any geometrical symmetry to go with the law of electric charge conservation. But not all symmetries are geometrical in nature. Take, for example, the phenomenon of inflation in economics. As the real value of the dollar declines, so the wealth of someone on a fixed income declines with it. If, however, a person has an index-linked income, their real earning power is independent of the value of the currency. We may say that index-linked income is symmetric under inflationary changes.

In physics, too, there are many examples of non-geometrical symmetries. One of these concerns the work necessary to lift a weight. The energy expended depends on the difference in height through which the weight is raised (it does not depend on the route taken). The energy is independent, however, of the absolute height: it would not matter whether the heights were measured from sea level or ground level, because it is only the height difference which is involved. There is a symmetry, therefore, under changes in the choice of zero height.

A similar symmetry exists for electric fields. In this *voltage* (electric potential) plays a role analogous to height. If an electric charge is moved from one point to another in an electric field, the energy expended depends only on the voltage *difference* between the end points of its path. If a constant extra voltage is applied to the whole system, the energy expended does not alter. Here is yet another hidden symmetry of Maxwell's electromagnetic equations!

The three examples given above illustrate what physicists call *gauge symmetries*. We can think of the symmetries involved as a 're-gauging' of money, height, and voltage respectively. All are abstract symmetries, in the sense that they are not geometrical in nature. We cannot look at them and see the symmetry. Nevertheless, they are still important indicators of the properties of the systems concerned. Indeed, it is precisely the gauge symmetry for voltages that ensures the conservation of electric charge.

The nuclear identity crisis

As a final example of an abstract symmetry which will turn out to be of major importance in later chapters, we shall turn to the strong nuclear force that acts between protons and neutrons. Experiments show that the strength and other properties of this force do not depend on whether the particles involved are protons or neutrons. In fact, the proton and the neutron are remarkably similar particles. Their masses lie within 0.1 per cent of each other. They have the same spin, and they respond identically to the nuclear force. The only distinguishing feature is the electric charge carried by the proton, but as far as nuclear forces are concerned the electric charge is of little importance and acts merely as a label. The charge serves to identify the proton and distinguish it from the neutron, but it plays no role in the nuclear force which binds the neutrons and protons together. Stripped of its charge, the proton suffers a desperate identity crisis.

The close similarity of the proton and the neutron suggests a symmetry at work. Nuclear processes would remain unchanged if, by some magic, we could swap the identities of all neutrons and protons. But one can go further than this. Imagine that we have a magic knob with a pointer on it. Rotating the knob is going to allow us to turn protons into neutrons. When the pointer is directed upwards, the particles concerned are protons; when the knob is turned around and the pointer is directed downwards, all the protons will have turned into neutrons (*see* Figure 8). This is a purely imaginary experiment, of course, since we cannot really change the identities of protons and neutrons. It is an abstract idea which will uncover an abstract symmetry, but nevertheless one that is of great value in helping us understand the nature of the nuclear force.

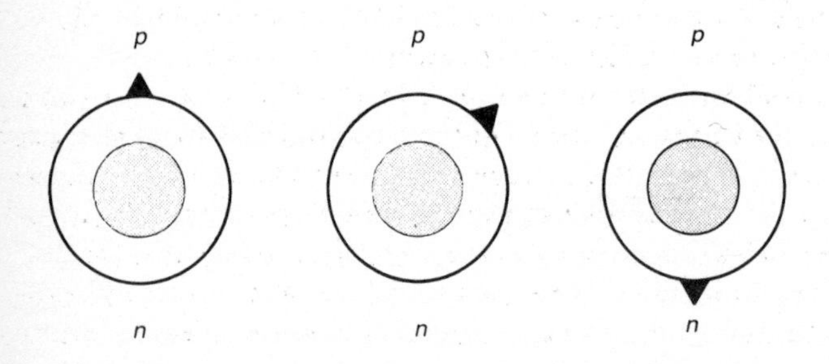

Figure 8. The magic knob. Rotating the knob gradually alters the identities of nuclear particles. With the pointer directed upwards, the particles are 100 per cent protons (p). As the knob is turned, so the particles acquire a hybrid identity, part-proton, part-neutron. With the pointer directed downwards, all the protons will have turned into pure neutrons (n). This process is entirely imaginary, but it depicts an underlying abstract symmetry of nuclear forces.

Suppose now that the change in identity from proton to neutron takes place not abruptly, but continuously as the knob is turned. At intermediate settings of the pointer the particles are neither pure protons nor pure neutrons, but a sort of hybrid of the two. As the pointer moves away from the twelve o'clock position so the 'proton-ness' of the particles begins to dwindle, while their 'neutron-ness' grows. You may find it hard to image 'proton-ness' and 'neutron-ness', or what it means for a particle to be a neutron–proton hybrid. An alternative interpretation of the intermediate settings of the pointer is that, on inspection, a given particle is sometimes found to be a proton and sometimes a neutron. The particle has a sort of identity crisis and flips randomly back and forth between the two possible states of 'proton' and 'neutron'. With the pointer near the twelve o'clock position, the particle spends most of its time as a proton, and so on inspection is most likely to be found as a proton. As the pointer moves closer to the six o'clock position, so the particle is more and more likely to be found to be a neutron. With the pointer directed exactly downwards, the proton content has faded completely, leaving 100 per cent neutron.

If the magic knob is equipped with both up and down pointers (Figure 9) we can envisage the same rotation of the knob that changes protons into neutrons as changing neutrons into protons. The knob set as shown in

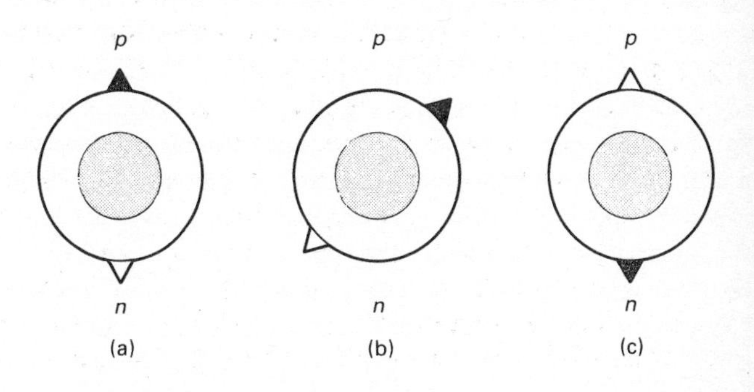

Figure 9. Equipped with two pointers, the magic knob metaphor can describe the transposition of protons and neutrons. The shaded pointer acts as in Figure 8, while the unshaded pointer describes the simultaneous convertion of neutrons into protons. Case (*a*) describes the population of protons and neutrons we actually observe. In (*b*) protons acquire an admixture of 'neutron-ness', while neutrons acquire an equal admixture of 'proton-ness'. With the knob set as in (*c*), all original protons have changed into neutrons, and all neutrons into protons. Because of an underlying symmetry, the nuclear forces are independent of the position of the knob.

Figure 9(*a*) corresponds to the present state of our universe. As it is rotated (Figure 9(*b*)), so all protons start turning into neutrons, while all neutrons begin changing into protons, and the admixture of neutron-ness among the protons is the same as the admixture of proton-ness among the neutrons. When the knob has been given a full half-turn, all protons and neutrons have swapped identity (Figure 9(*c*)).

The knob and pointer imagery are helpful in describing the nature of the symmetry which underlies nuclear forces. Using this language we can say that, in essence, the nuclear forces are *independent of the setting of the pointer*. It can be up, down, sideways, or at any intermediate angle. The nuclear forces do not change. It is a property known by the rather formidable title of 'isotopic spin symmetry'. The 'isotope' part refers to the fact that nuclei which differ in the numbers of neutrons are called isotopes, while the symmetry properties involved here are closely analogous to those of intrinsic spin mentioned in Chapter 2.

Physics and fantasy

The concept of isotopic spin is a marvellous example of the power of abstract reasoning in physics, and one which will turn out to have deep implications, as we shall see. In the real world there are no 'magic knobs and pointers', no devices for mixing the identities of protons and neutrons. The idea is a purely theoretical construct, a fantasy. Nevertheless, it is logically possible. We can imagine doing such a thing even though in the real world it is forbidden. The fact that an imaginary process can have a fundamental bearing on real physics in the real world may come as a surprise, but pursuing such an approach is an important tool in the armoury of the modern physicist. He sees physics as a model which describes the real world of experience. The model may contain many auxiliary features which are themselves not part of that real experience, but which nevertheless play a vital role in the theory.

Why should physicists invent totally imaginary, abstract entities to model the real world? Is it not possible to deal entirely with quantities that can actually be observed? After all, theories can be checked only by concrete observation, and the imaginary features of the model will never show up explicitly in any predictions of the theory dealing with actual observations. So why include them?

The incorporation of imaginary elements into physical theories is one of the most difficult practices for a professional physicist to justify to the layman. Of course, if a particular feature, such as isotopic spin symmetry, renders the model a brilliant success, then the physicist can simply reply, 'I put it in because it works!'

What can appear baffling is how the physicist knows just which piece of Alice-in-Wonderland abstraction should be included. In view of the fact that the feature concerned is purely imaginary, it might appear to a cynic that anything goes: 'No need to pick things that are actually there in the world outside; choose what you want from your imagination.' The range of choice is infinite. How do you pick the 'right one'?

It is at this stage that professional physicists start using words like beauty, mathematical elegance, and symmetry. Although the inclusion of imaginary or abstract ideas, such as gauge symmetry, is not logically necessary to make a successful theory – in principle all theories could be formulated entirely in terms of observable quantities – nevertheless a theory may be vastly simpler and more appealing if such abstract elements are included.

Consider the idea of a field, which has proved so successful in physics and engineering. The field concept was introduced by Faraday and Maxwell as an abstraction. We can't directly see or touch an electromagnetic field. We only know it is there by the action it has on electric charges. On the other hand, if the field is only ever produced by yet more electric charges, what we are really dealing with is an interaction between electric charges. If the charges themselves are in practice the entities that you observe, why do we need a field at all? Why not just talk about the way charges act on each other across space and formulate all equations of the theories of electricity in terms of them alone?

This can be done. The difference is that the theory which results is messy and complicated in an undefinable way, yet in a way that is immediately recognized by a professional physicist. Field theory is so much more 'elegant'. The mathematics is smoother, more harmonious and interlocking, more economical. And more *suggestive*.

This is an important point. Often the elegant, streamlined, abstract theory points the way to new developments in physics that simply would not have presented themselves if the theorists had stuck with models formulated entirely in terms of concrete, observable quantities. For example, the quantum theory of fields, so vital to the superforce and other recent developments in fundamental research, could never have emerged had the field concept not become current among physicists.

When an abstract concept becomes so successful that it permeates through to the general public, the distinction between real and imaginary becomes blurred. The physicist's imaginary quality becomes invested with a familiarity that seems to turn it into a real thing. This is what has happened in the case of energy. The concept of energy was introduced into physics as an abstract idea. What made it appealing was the law that energy is always conserved, never created or destroyed. Yet what *is* energy? Can you see it or touch it?

When a weight is lifted above the ground, some work has to be done to raise it. We say that energy has been expended, but the law of energy conservation ensures us that the energy is still there somewhere. We can see the straining muscles of the weight-lifter. We can imagine that we are actually observing energy in action in his contorted features and bulging biceps. But when the weight is raised, and perhaps laid comfortably on a platform, where has the energy gone? Can we still see it?

The physicist says that the energy is *stored* in the weight, by virtue of its elevated location. This is the elusive concept of 'potential' energy. The energy is there all right, albeit invisible, and could be recovered easily enough by the simple expedient of taking the support away and letting the weight crash down. The crashing sound is some of that stored energy being released.

Energy is thus an imaginary, abstract concept which nevertheless has become so much a part of our everyday vocabulary that we imbue it with concrete existence. 'I haven't enough energy to dig the garden' is not a statement likely to attract stares of incomprehension. Nobody asks what colour your energy is, or demands that you pour it into a bowl so that its volume can be measured. Yet it is accepted without question that you have energy just as you have skin and bones.

Energy is one of the physicist's more enduring abstract concepts. It is of enormous assistance in describing a wide range of physical processes. The law of energy conservation embodies a huge variety of experiences that, in the absence of the energy concept, would all have to be discussed separately. Energy lets us connect many ideas together, and as such it can be deemed beautiful.

And therein lies its appeal and utility. Nature *is* beautiful. We don't know why this is so, but experience teaches us that beauty implies utility. Successful theories are always beautiful theories. They are beautiful not *because* they are successful, but because of their inherent symmetry and mathematical economy. Beauty in physics is a value judgement involving professional intuition and cannot readily be communicated to the layman, because it is expressed in a language that the layman has not learned, the language of mathematics. But to one who is conversant with that language, the beauty is as apparent as poetry.

This brings me back to where I came in. Mathematics is a language, the language of nature. If you can't speak a language you can't understand the beauty of its poetry. There are always sceptics who say, 'What is this mysterious mathematical beauty you speak of? I don't see anything beautiful about a mess of symbols. You physicists are just deluding yourselves.' I like to reply by comparing mathematics with music. For someone who had heard only single musical notes, the beauty of a symphony would be impossible to explain. Yet who would deny that there is real beauty in a symphony, albeit of an abstract and indefinable nature? Likewise, for a person whose experience of mathematics is limited to counting numbers, how can one communicate the sense of delight, the deep and meaningful appeal, of Maxwell's equations? Nevertheless, the aesthetic quality is there sure enough. And physicists of good mathematical taste produce altogether better theories than Philistines, just as do their counterparts in musical composition.

It is one of the great tragedies of our society that from fear, poor teaching, or lack of motivation the vast majority of people have shut themselves off from the mathematical poetry and music of nature. The sweeping vista that mathematics reveals is denied to them. They may delight over the scent of a rose or the colour of a sunset, but a whole dimension of aesthetic experience is foreclosed to them.

5
The Four Forces

The source of all change

Ever since man has contemplated the world about him he has recognized the existence of change. The world is full of activity – the motion of the sun, the blowing wind, the soaring birds, the running stream. Ancient man saw the seasons pass, people grow old, his primitive tools wear out.

But what was the reason for change and motion? Some things, such as animals, seemed to contain their motive power within themselves, whereas others – rocks, arrows, axes – clearly needed external propulsion to make them move. At first, no clear distinction was made between an object moving through space and change of any general sort. Precise ideas of velocity and acceleration were not well formulated. Our remote ancestors certainly conceived of forces that shaped the world and brought about change, but these forces were magical in quality and inseparable from their beliefs in the gods and demons who ruled their universe.

The Greek philosophers adopted a more systematic study of change and motion, but still did not fully comprehend the causes. Aristotle believed that the key to understanding motion was resistance. He observed that a body moves more freely, hence faster, through a tenuous medium such as air than through a dense medium such as water; in both cases a motive power is needed to overcome the resistance of these fluids. He poured scorn on the atomist idea of particles in free motion in a void, because a void, being empty of substance, could offer no resistance. The particles would therefore move with infinite speed, a nonsensical prospect.

The modern engineer's concept of a force was not fully developed until the seventeenth century, which saw the acceptance of Newton's laws of mechanics. Newton's masterstroke was to realize that motion as such did not necessarily require a force. A material body will move at uniform speed in a fixed direction without any external agency to push or pull it. Only deviations from uniform motion require explanation, i.e. the presence of forces. According to Newton, forces produce accelerations, and he provided a precise mathematical formula linking the two.

Newton's theory immediately explained a puzzle about the Earth's motion around the sun. No visible agency exists to push or pull the Earth along the line of its orbit. In Newton's theory, none is needed. The fact of the Earth's motion is not an issue; it is its departure from uniformity (straight line motion at constant speed) that requires explanation. The Earth's path through space curves around the line of the direction towards the sun, a fact readily explained by the sun's gravitational force.

Newton's mechanics rapidly became established as a successful description of force and motion, and now forms the basis of all engineering. It makes no reference, however, to the origin of the forces which accelerate matter. At first sight these forces seem to be many and varied: the impact of the wind or a running stream, the pressure of air or water, the insistent push of expanding metal, the violent outburst of exploding chemicals, the tug of stretched elastic, human muscle power, the weight of heavy objects, and so on. Some forces seem to act directly by contact with a body, as in the pull of a rope, while others, like gravity, appear to act at a distance across empty space.

Despite this great variety, careful study has shown that all of nature's activity can be reduced to the operation of just four fundamental forces. These forces are ultimately responsible for all the activity of the world; they are the source of all change. Each force has its similarities to and differences from the others. Understanding the properties of these four forces is a major task for the physicist, and constitutes an essential preliminary on the road to the superforce.

Gravity

Historically, gravity was the first of the four forces to be treated scientifically. Although man was always aware of gravity, and based the very notions of up and down on it, the true role of gravity as a force of nature was not fully recognized until Newton's theory of gravitation was published in the seventeenth century. Until then, gravity was inextricably linked to the Earth, and mixed up with whatever cosmological beliefs held sway. Aristotle, who believed the Earth to lie at the centre of the universe, regarded the tendency of bodies to fall to the ground as simply an instance of a general principle that all bodies have a 'natural place' in the universe, and tend towards it. Corporeal bodies therefore tended downwards, while gaseous substances tended heavenwards, towards the less substantial spiritual realm. The ethereal elements of the heavens themselves revolved around the Earth on exactly circular paths, this being geometrically the perfect motion.

With the establishment of more modern astronomical ideas in the Middle Ages, it was recognized that gravitational forces are not limited to the Earth, but act between the sun, moon, and planets and, indeed, all bodies in space. One of the most convincing demonstrations of this fact was Newton's explanation for the ocean tides in terms of the moon's gravitational pull. Newton's inverse square law embodied the 'long-range' nature of gravity. By this it is meant that although the force of gravity dwindles with distance, its effects can still reach out across space and be felt a long way away. That is fortunate, because gravity literally holds the universe together: it grips the planets in their orbits around the sun, it binds the stars to the galaxy, and it also prevents the stars from boiling away into the vacuum of space. In fact, on an astronomical scale, gravity is generally the dominant force.

An important feature of gravity is its universality. Nothing in the cosmos escapes its grip. Every particle is subject to gravity, or 'couples' to gravity, to use the physicist's jargon. Even energy responds to gravity. Likewise, every particle is a source of gravity. Furthermore, the strength with which particles couple to gravity is always the same, a fact implicit in the famous observation attributed to Galileo that all bodies fall equally fast whatever their weight or constitution.

The force of gravity between particles is always attractive; it acts to draw them together. Repulsive gravity, or 'antigravity' as it is sometimes called, has never been observed. The reason for this is well understood. Gravitational repulsion would require negative energy. As the energy locked up in a particle is always positive, giving it a positive mass, particles always try to gravitate towards each other. Negative energy sounds incomprehensible. However, although particles cannot possess negative energy, it is possible for a field to do so, with profound consequences, to be explored in a later chapter.

Perhaps the most surprising thing about gravity is its extreme feebleness. The force of gravity between the components of a hydrogen atom is 10^{-39} of the electric force. If a hydrogen atom were bound by gravity rather than electricity the smallest electron orbit would be larger than the observable universe! In fact, at the level of individual subatomic particles, gravity is so weak that physicists have been inclined to ignore it entirely. It never features in any of the particle processes so far observed.

Even at the level of macroscopic objects, the effects of their own gravity escape our attention. When you walk down a street, the large buildings exert minute gravitational tugs, but they are too small to be felt. Delicate equipment can, however, respond to these forces. As early as 1774 a Scot, Nevil Maskelyne, observed the tiny deflection of a plumb-line from the vertical caused by the gravitational force of a nearby mountain. In 1797 Henry Cavendish performed a famous experiment in which he very carefully measured the miniscule force of attraction between two small balls attached to the ends of a horizontally suspended wooden rod, and two large lead spheres. This was the first time a gravitational force between two bodies had been observed in the laboratory.

It may seem surprising that if gravity is so weak we feel it at all. How can it be the dominant cosmic force? The answer lies in its universality. Because every particle of matter gravitates, and always attractively, gravity accumulates as more and more matter is aggregated. You feel gravity in daily life because every atom in the Earth is pulling on you in concert. The effect of any single electron or proton is utterly negligible, but when they all pull together the result can be a substantial force. If there were as many antigravitating particles as gravitating ones, they would tend to neutralize each other and the force of gravity, whilst still existing, would never be noticed; it would be too weak to manifest itself.

Gravity can only be described properly as a *field*. We can envisage every particle as the source of a gravitational field which emanates from the particle, surrounding it with an invisible halo of influence. Another particle immersed in this gravitational field feels a force. The field is more than just a way of speaking about gravity though. As mentioned in Chapter 2, it can support wavelike disturbances. Just as Maxwell discovered that waves could be set up in the electromagnetic field and travel through space, so Einstein found that waves can be set up in the gravitational field.

Although Newton's theory of gravitation was perfectly adequate for over 200 years, it became a casualty of the new physics which erupted in the first decades of the twentieth century. A long-standing discrepancy of Newton's theory concerned the orbit of the planet Mercury, which is not quite elliptical. A small twisting, or precessing, of the orbit is caused by gravitational disturbances of the other planets, but when the effect was taken into account there remained a residual twist of a mere 43 seconds of arc per century which could not be accounted for by Newton's theory.

More seriously, Newton's law of gravitation clashed with the emerging theory of relativity. According to Newton, the gravitational force between two bodies is transmitted instantaneously across space, so that if the sun, by some magic, were to disappear suddenly, the Earth's orbit would instantly cease to curve, even though we should not see the sun's disappearance for the eight minutes that it takes for sunlight to reach us. Einstein's theory of relativity forbids any physical influence from travelling faster than light, and clearly comes into conflict with Newton's gravitational theory.

It was his attempt to generalize the theory of relativity to include gravitation that led Einstein to his (1915) 'general theory of relativity', which not only displaced Newton's gravitational equation, but changed the entire conceptual basis for our understanding of gravitation. In Einstein's theory, gravity is not really a force at all, but a manifestation of the curvature, or warping, of spacetime. Bodies are not 'forced' into curved orbits by gravity, they simply follow the straightest, easiest path through a curved spacetime. According to Einstein, gravity is just geometry.

Newton's theory remains satisfactory for all practical purposes, such as aircraft and spacecraft navigation, and is still adequate for the description of most astronomical systems. It fails, however, when gravitational fields become intense, as they do near collapsed objects like neutron stars or black holes. Even for moderate gravitational fields, the effects of curved spacetime can still be detected. The twisting of Mercury's orbit, for example, is a consequence of the sun's spacewarp. Also, as mentioned in Chapter 2, very sensitive clocks can detect timewarps at the Earth's surface.

Electromagnetism

Though gravity was the first force to be properly understood in the scientific sense, electromagnetism has been equally familiar to people from time immemorial. Electric forces are conspicuously unleashed during lightning strikes, and can be seen at work during corona discharges and other luminous atmospheric phenomena. Magnetic forces are responsible for the complex patterns observed in auroral displays.

The Greek philosopher Thales is credited with the first definitive identification of electricity. He found that if a piece of amber was rubbed it acquired the ability to pick up small objects. The word *elektron* is Greek for amber. This curious phenomenon was further studied in the Middle

Ages by William Gilbert, a physician to Queen Elizabeth I, who found that many other substances shared the electric property of amber. Further investigation in England and continental Europe established the fact that certain substances act as insulators. The French scientist Charles Dufay discovered that electric charge comes in two varieties, which we now call positive and negative.

During the eighteenth and early nineteenth centuries, electricity came to be understood in more detail following experiments by Benjamin Franklin and Michael Faraday. It was known that like electric charges repel but unlike charges attract, in both cases with a strength that is governed by a simple mathematical formula: the electric forces diminish with distance according to the same 'inverse square' relation that Newton had earlier deduced for gravity. However, electric forces are vastly stronger than gravitational forces. In contrast to the minute gravitational forces that Cavendish had detected with special equipment, electric forces between everyday-sized objects are readily observed.

Faraday's work suggested that electricity was present in the atom, but it was not until J. J. Thompson discovered 'cathode rays' in the 1890s that the existence of the electron was firmly established. Today, we know that electric charge is always attached to particles of matter in exact multiples of a fundamental unit, a sort of 'atom' of charge. Why this should be so is an interesting problem. Not all particles, however, carry electric charge. The photon and the neutrino, for example, are electrically neutral. In this respect electricity differs from gravity. All particles of matter 'couple' to the gravitational field, but only charged particles couple to the electromagnetic field.

As with electricity, natural magnetism was also identified by the early Greeks. By about 600 B.C. they were familiar with the properties of lodestones (iron oxide) which they found could exert an influence on each other even at a distance. About 500 years later the Chinese discovered the intriguing directional properties of lodestone and constructed the first crude form of compass. Its use was reserved, however, for mystical purposes, and it was several centuries before compasses became navigational aids.

Towards the end of the sixteenth century European scientists began to appreciate the true nature of magnetism. Gilbert demonstrated that the Earth itself behaved like a magnet with properties that closely resembled those of a model lodestone sphere which he had constructed. Magnetism was also found to come in two varieties, called north and south poles after the Earth's magnetism. As with electricity, like poles repel while unlike poles attract. Unlike electricity, however, magnetic poles always seem to occur in pairs, north and south. In a typical bar magnet one end will act as a north pole, the other end as a south pole. If the bar is cut in half new poles appear at the position of the cut, producing two magnets each with a north and south pole. However hard one tries, it is impossible to isolate a single magnetic pole, or *monopole*, this way. Could it be that isolated magnetic poles are forbidden in nature? If so, why? We shall see that a study of the superforce provides answers to these fascinating questions.

The force between magnetic poles also obeys an inverse square law, as with electricity and gravity. Thus, electric and magnetic forces are 'long-ranged', and can be discerned over large distances. For example, the magnetic field of the Earth extends far out into space. The sun also has a magnetic field which permeates the whole solar system. There is even a galactic magnetic field.

In the early nineteenth century a deep connection between electricity and magnetism was found. In Denmark Hans Christian Oersted discovered that an electric current generates a magnetic field around itself, while Faraday showed that a changing magnetic field induces an electric current to flow. These discoveries formed the basis for the electric dynamo and the generator which today play such an important role in engineering.

As already recounted in Chapter 4, the decisive step was taken by Maxwell in the 1850s when he united electricity and magnetism into a single theory of electromagnetism, the first unified field theory. With appropriate refinements to take into account quantum effects, Maxwell's theory remained intact, brilliantly successful, until 1967, when the next great step in unification took place.

The weak force

Though it was not appreciated at the time, mankind witnessed the weak force in action as long ago as 1054, when oriental astronomers spotted the sudden appearance of an intensely bright star in a region of the sky where none had been seen before. The 'guest star' burned brightly for several weeks, rivalling even the planets in brilliance, before slowly fading to obscurity. Today's astronomers recognize the 1054 outburst as a supernova explosion, i.e. the cataclysmic disintegration of an aging star occasioned by the abrupt collapse of its core and the attendant release of a huge pulse of neutrinos. Armed only with the weak force, these neutrinos blast the outer layers of the star into space, producing a ragged cloud of expanding gas. The 1054 supernova remains visible as a fuzzy patch of nebulosity in the constellation of Taurus.

Supernovae provide one of the few instances when the weak force manifests itself conspicuously. It is by far the weakest force after gravity, and in many systems where it is present its effects are swamped by the electromagnetic force or the strong force.

The existence of a weak force dawned only slowly on the scientific community. The story began in 1896 when Henri Becquerel accidentally discovered radioactivity while investigating the mysterious fogging of a photographic plate that had been left in a drawer next to some uranium sulphate crystals. A systematic study of radioactive emissions was undertaken by Ernest Rutherford, who demonstrated that two distinct types of particles were emanating from the radioactive atoms. These he called alpha and beta. The alphas were heavy, positively charged particles which turned out to be fast-moving helium nuclei. The betas were shown to be high-speed electrons.

The details of beta radioactivity were not fully understood until the 1930s. The process was peculiar. At first sight it appeared that one of the fundamental laws of physics, the law of energy conservation, was being violated. Some energy seemed to be missing. Wolfgang Pauli rescued the law by suggesting that another particle – a neutral, intensely penetrating particle that nobody had yet detected – was coming out with the electron. Enrico Fermi dubbed the unseen particle the 'neutrino', meaning 'little neutral one'. It turned out to be so elusive that neutrinos were not definitively spotted until the 1950s.

Still there was a mystery. The electrons and neutrinos were emanating from unstable nuclei. But physicists had irrefutable evidence that neither of these particles existed *inside* nuclei. So where did they originate? Fermi proposed that the electrons and neutrinos did not exist prior to their ejection, but that they were instantaneously created in some way out of the energy present in the radioactive nucleus. The quantum theory had already shown that the emission and absorption of light could be understood as the creation and destruction of photons; Fermi hypothesized that the same sort of thing could happen to electrons and neutrinos.

Confirmation of Fermi's proposal came from the behaviour of free neutrons. Left to themselves, neutrons disintegrate after several minutes, leaving a proton, an electron, and a neutrino. One particle disappears and three new particles appear. It soon became clear that known forces could not explode a neutron in that way. Some other force had to be driving beta decay. Measurements of the rate of such decays established that the force concerned was extraordinarily weak, much weaker than electromagnetism (though still immensely more powerful than gravity). The need for a new 'weak force' was finally recognized.

With the discovery of unstable subnuclear particles, physicists found that the weak force was responsible for many other transmutations. Most known particles, in fact, 'couple' to the weak force. For the ghostly neutrino, its weak action (leaving aside gravity) is the only way it manifests its existence in the world.

The weak force is quite different in character from either gravitational or electromagnetic forces. For a start, it does not exert a push or pull in the engineering sense, except in events like supernova explosions. Instead, it is restricted to driving transmutations in the identity of particles, often propelling the products to high speed. Second, the activities of the weak force are restricted to an extremely limited region of space. Indeed, only since the early 1980s has it been possible to accurately measure the range of the weak force. For a long while it seemed that its action was essentially point-like, occupying a region of space too small to discern. In contrast to the 'long-range' nature of gravity and electromagnetism, the weak force is inoperative beyond about 10^{-16} cm of its source. It cannot, therefore, act on macroscopic objects, but is confined to individual subatomic particles.

Though the theory of the weak force developed by Fermi and others in the 1930s was steadily improved over the years, deep inconsistencies remained, and it became clear that a proper understanding of the weak force had not been attained. A new theory, drawing upon Fermi's basic ideas but adding some crucial new features, was worked out in the late 1960s by Steven Weinberg, then working at Harvard University, and Abdus Salam of Imperial College, London. This step represented the biggest advance on the road to the superforce since Maxwell developed his electromagnetic theory, and will be described in detail in Chapter 8.

The strong force

The existence of a strong force slowly dawned on physicists as the structure of the atomic nucleus became clear. Something had to bind the protons together against the repulsion caused by their electric charge. Gravity is far too weak to achieve this, and so a new type of force was clearly necessary, a very strong force, stronger than electromagnetism. No trace of this strong nuclear attractive force can be discerned outside the confines of the nucleus, and so the new force had to be very short in range. In fact, it fades away rapidly beyond a distance of about 10^{-13} cm from a proton or neutron. Consequently, although it is the strongest of nature's four forces, the strong force cannot be directly discerned in macroscopic bodies.

Both neutrons and protons are subject to the strong force, but electrons are not. Nor are neutrinos or photons. Generally, only the heavier particles couple to the strong force. Its effects manifest themselves both as a conventional 'pulling' force that holds the nucleus together and also, like the weak force, in causing some unstable particles to decay. Because of its strength, the strong nuclear force is the source of great energy. Perhaps the most important example of energy which is released by the strong force is sunlight. The cores of the sun and other stars are nuclear fusion reactors under the control of the strong force. It is also this force which liberates the energy of a nuclear bomb.

Early attempts to understand the strong force met with only limited success. No simple mathematical description seemed completely satisfactory. The force did not seem to vary with distance in any very straightforward manner, and nuclear physicists who wanted to model it were compelled to introduce many arbitrary parameters. It was almost as though the strong force was an amalgam of lots of forces with differing properties.

While they were grappling with this complication, the quark theory of nuclear matter was proposed, in the early 1960s. This theory recognizes that neutrons and protons are not elementary particles but composite bodies consisting of the union of three quarks each. Obviously, some sort of force was needed to bind or 'glue' the quarks into a trio, and it became clear that, according to the quark theory, the overall force between whole neutrons and protons must be merely a residue of the more powerful inter-quark force. The reason why the strong force appears so complicated was now apparent. When a proton sticks to a neutron or another proton, the adhesion involves a total of six quarks, each of which interacts with all the others. Most of the force goes into gluing the quark trios firmly together, but a little is left over to stick the two sets of three to each other.

Once its true nature as an inter-quark force was appreciated, the strong force became much easier to model mathematically. In later chapters we shall see how this description brings the strong force into parallel with the other forces, and points the way to the existence of a unifying superforce.

6
The World of Subatomic Particles

Atom-smashing

It is often said that there are two types of science, big science and little science. Atom-smashing is big science. It uses big machines and big budgets, and nets the lion's share of Nobel prizes.

Why do physicists want to smash atoms? The simple answer – to find out what is inside them – has an element of truth in it, but there is a more general reason. Atom-smashing is really a misnomer. High-energy particle collisions would be a more accurate description. When subatomic particles collide at great speed, the shock of impact bursts open a new world of forces and fields. The fragments of energized matter which erupt from these encounters carry secrets about the workings of nature that have lain unsuspected since the creation, buried in the deepest recesses of the atom.

The machines which bring about these collisions are the particle accelerators, awesome in size and cost. Their dimensions are measured in miles, and they dwarf even the laboratories where the collisions are studied. In other scientific enterprises the equipment is contained in the laboratory. In high-energy particle physics the laboratories are grafted on to the equipment. Recently, the Centre for European Nuclear Research (CERN) near Geneva embarked on the construction of a several hundred million dollar, ring-shaped machine to be housed in a circular tunnel 27 km in circumference. Called LEP (large electron–positron ring), it is designed to accelerate electrons and their antiparticles (positrons) to within a whisker of the speed of light. To get some idea of the colossal energy involved, imagine that instead of electrons, a penny piece were being accelerated. At the end of the run, the coin would have enough energy to produce 100 million million dollars worth of electricity! Small wonder that this sort of enterprise is referred to as 'high-energy' physics. Travelling in opposite directions around the ring, the electron and positron beams will be allowed to collide head on, annihilating each other and liberating enough energy to create dozens of other particles.

What are these particles? Some of them are the components that go to build up you and me; the protons and neutrons of atomic nuclei, and the electrons which orbit around them. Others are not normally found in ordinary matter; they exist only fleetingly before decaying into more familiar forms. The number of different species of these unstable, temporary particles is astonishing: several hundred in total have been catalogued so far. Like the stars, the unstable particles are too numerous to qualify for names. Many are known by Greek letters only, some by numbers.

It is important to understand that these many different unstable particles are not in any straightforward sense the *constituents* of protons, neutrons, or electrons. When electrons and positrons collide at high energy they do not 'break open' and spill out a shower of subatomic debris. Even high-speed collisions involving protons, which certainly do have objects (the quarks) inside them, do not actually involve the protons being smashed apart in the usual sense. It is better to envisage the debris which emerges from these collisions as being created 'on site' out of the energy of impact.

Twenty years ago physicists were completely bewildered by the number and variety of subatomic particles that were being discovered in seemingly endless succession. They wondered what all the particles were *for*. Were they simply like animals in a zoo, with vague family resemblances but no systematic relationships connecting them? Or were they, as some optimists believed, the key to the universe? Were the particle physicists just producing insignificant random chippings of matter, or was a pattern emerging, a dimly perceived order that hinted at a rich and elaborate structure in the subnuclear world? Today, there is no longer any doubt. There is a deep and meaningful order in the microcosmos, and though as yet we can glimpse this order only in outline, we are beginning to understand what all these particles mean.

The first step in this understanding came from simple systematic cataloguing, rather like the eighteenth century biologists carefully listed the plant and animal species. The vital statistics of subatomic particles are mass, electric charge, and spin.

Mass is roughly the same as weight, and particles with a lot of mass are often referred to as 'heavy'. Einstein's $E = mc^2$ relation tells us that the mass of a particle depends on its energy and hence its speed. A particle is heavier when in rapid motion than at rest. The mass of interest is the *rest* mass, as this cannot be varied. If a particle has zero rest mass then it moves at the speed of light. The photon is one obvious example of a zero rest-mass particle. Of the particles with non-zero rest mass, the electron is believed to be the lightest. The proton and neutron are nearly 2 000 times heavier, while the heaviest particle so far produced in the laboratory (the Z) is about 200 000 times heavier than the electron.

Electric charge does not come in a wide range of values but, as we have seen, is always found in fixed multiples of a fundamental unit. Some particles, like the photon or the neutrino, have no electric charge. If the positively charged proton is designated as having +1 units, then the electron has exactly −1 units.

The subject of spin was introduced in Chapter 2. It, too, always comes in fixed multiples of a fundamental unit which, for historical reasons, is taken to be ½. Thus the proton, neutron, and electron have spin ½ and the photon has spin 1. Particles are also known with spin 0, 3/2, and 2. No fundamental particle has a spin greater than 2 and theorists believe such an entity is impossible.

The spin of a particle is a vital indicator of its nature, and all particles fall into one of two quite distinct classes. Those with spin 0, 1, or 2 are called 'bosons' after the Indian physicist Satyendra Bose, while half-integral spin particles (those with spin ½ or 3/2) are called fermions, after Enrico Fermi. Which of these two classes a particle belongs to is perhaps its most vital vital statistic.

Another important statistic is the lifetime of the particle. Until recently it was believed that electrons, protons, photons, and neutrinos are absolutely stable, i.e. that they have an infinite lifetime. The neutron can remain stable when trapped in a nucleus, but a free neutron decays in about 15 min. All other known particles are highly unstable, their lifetimes varying from microseconds down to 10^{-23} s. If these times seem incomprehensibly short, remember that a particle travelling at near the speed of light (which most accelerator products do) can cover 300 m in a microsecond.

Particles that decay do so by the action of quantum processes, and so there is always an element of unpredictability involved. The lifetime of a given particle cannot be foretold. What can be predicted is the average lifetime, on a statistical basis. Usually quoted as the 'half-life' of a particle, which is the time required for a population of identical particles to become depleted by 50 per cent. Experiment shows that the decline in population is 'exponential' (*see* p. 55), for which the half-life is 0.693 times the average life.

It is not enough to know that a particle exists; physicists also want to know what it does. This is in part determined by the quantities listed above, but also by the type of forces which act on it and within it. The factor which determines above all else the properties of a particle is whether or not it feels the strong force. Particles which 'couple' to the strong force stand in a class apart, and are given the name *hadrons*. Others, which feel the weak but not the strong force, are called *leptons*, meaning 'light thing'. We shall take a brief look at each of these families in turn.

Leptons

The best-known lepton is the electron. Like all leptons, it appears to be an elementary, point-like object. As far as we can tell, an electron has no internal structure, i.e. it is not 'built out of' anything. Though leptons come both with and without electric charge, they all have spin ½, so they are all fermions.

Another well-known lepton, this time of the chargeless variety, is the neutrino. As explained in Chapter 2, neutrinos are elusive to the point of ghostliness. Because they feel neither the strong force nor the electromagnetic force, they are almost completely oblivious to matter and pass right through it as if it were not there. Historically this made their very existence extremely hard to verify. It was more than three decades after neutrinos were first predicted to exist that they were finally detected in the laboratory. Physicists had to await the advent of nuclear reactors, which release prodigious quantities of neutrinos, before they could stop one in a head-on collision with a nucleus, and thus really prove its existence. Today, more controlled experiments can be performed with carefully regulated neutrino beams produced by decaying particles in an accelerator. The vast majority of neutrinos ignore the target completely, but the occasional one can be made to react and provide useful information about the structure of other particles and the nature of the weak force. Of course, experiments with neutrinos, unlike those with all other subatomic particles, do not require any protective shielding. Their penetrating power is so enormous they are completely harmless, and pass right through you without the slightest damage.

In spite of their intangibility, neutrinos enjoy a status unmatched by any other known particle, for they are actually the most common objects in the universe, outnumbering electrons or protons by a thousand million to one. In fact, the universe is really a sea of neutrinos, punctuated only rarely by impurities such as atoms. It is even possible that neutrinos collectively outweigh the stars, and therefore dominate the gravity of the cosmos. A report from a group of Russian scientists claims that neutrinos actually possess a tiny mass, less than one ten thousandth of the electron's mass, but enough, if true, to gravitationally overwhelm the universe and cause it to collapse at some time in the future. Therefore, although superficially it is the most harmless and ephemeral particle, the lowly neutrino could hold the power of total cosmic annihilation.

Of the remaining leptons, the muon was discovered in 1936 amid the products of cosmic rays, and it was among the first known unstable, subatomic particles. In all respects other than stability the muon is like a big brother of the electron; it has the same charge, spin, and response to forces, but a much bigger mass. It decays in about two-millionths of a second into an electron and two neutrinos. Muons are actually rather common particles in nature, and are responsible for much of the background cosmic radiation that can be detected at the Earth's surface by a Geiger counter.

For many years the electron and the muon were the only known charged leptons. Then, in the late 1970s, a third one was found, called the tauon. Weighing in at 3 500 electron masses, the tauon is definitely the heavyweight of the charged lepton trio, but aside from its mass it behaves just like an electron or a muon.

This does not exhaust the list of known leptons. In the 1960s it was found that there is actually more than one sort of neutrino. First, there is the type that gets created along with the electron when a neutron decays, and then there is another sort that turns up when a muon is created. Each type of neutrino goes hand in hand with its own charged lepton; hence there is the 'electron–neutrino' and the 'muon–neutrino'. It seems inevitable that

Table 1

Name	*Symbol*	*Mass*	*Charge*
Electron	e^-	1	−1
Muon	μ^-	206.7	−1
Tauon	τ^-	3536.	−1
Electron-neutrino	ν_e	0	0
Muon-neutrino	ν_μ	0	0
Tauon-neutrino	ν_τ	0	0

The six leptons come in both charged and neutral varieties (antiparticles are not included). The mass and charge are expressed in units of the electron's mass and charge. Some evidence exists that neutrinos may possess a minute mass.

there must be a third type of neutrino to go with the tauon, making three different species of neutrinos in all, and a grand total of six leptons (*see* Table 1). Of course, each sort of lepton has its own antiparticle as well, making twelve different leptons altogether.

Hadrons

In contrast to the handful of known leptons, there are literally hundreds of hadrons. This fact alone suggests that hadrons are not elementary particles, but composites of smaller objects. All hadrons feel the strong and weak forces and gravity, but they come in both electrically charged and neutral varieties. The best-known and commonest are the neutron and proton. All the others are very short-lived, and decay either in less than a millionth of a second under the action of the weak force, or much more rapidly (typically 10^{-23} s) from the effects of the strong force.

In the 1950s physicists were utterly bewildered by the number and variety of hadrons. Bit by bit they began to organize the data in a meaningful way according to the particles' vital statistics of mass, charge, and spin. Slowly a semblance of order began to emerge; patterns became apparent. There were hints of symmetries at work beneath the surface muddle of the data. The decisive step in unravelling the hadron mystery came in 1963 when Murray Gell-Mann and George Zweig of Caltech invented the quark theory.

The essential idea is very simple. Inside all hadrons are smaller particles called quarks. The rules say that quarks are permitted to stick together in one of two possible ways, either in trios or in quark–antiquark pairs. Combinations of three obviously produce heavier particles, and these are called *baryons*, meaning 'heavy ones'. The best-known baryons are the neutron and proton. Quark–antiquark pairs are somewhat lighter, and make up particles known as *mesons*. The name derives from the fact that the first mesons were intermediate in mass between electrons and protons. To account for all the then-known hadrons, Gell-Mann and Zweig introduced three distinct types or 'flavours' of quark, whimsically called 'up' (u), 'down' (d), and 'strange' (s). The various combinations of these flavours then explained the existence of the many different sorts of hadrons. Thus, a proton consists of two ups and a down quark, while a neutron contains two downs and an up (*see* Figure 10).

For the scheme to work properly it is necessary to suppose that quarks carry fractional electric charge. That is, they possess a quantity of charge

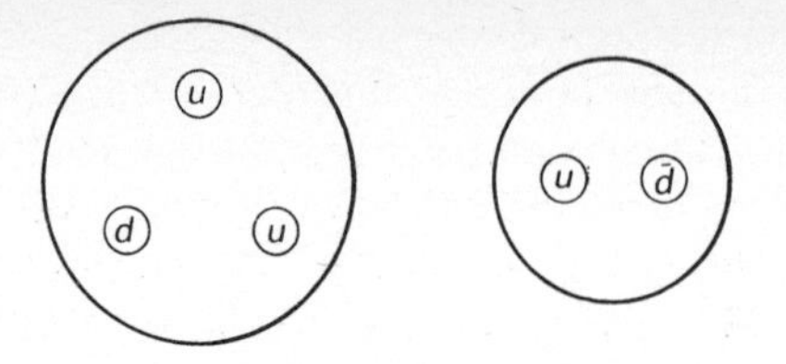

Figure 10. Hadrons are made of quarks. A proton (left) consists of two up (u) and one down (d) quark. The lighter pion (right) is a meson, and contains an up quark (u) and a down antiquark (d̄). Other quark combinations make up the remaining members of the hadron family.

that is either ⅓ or ⅔ of the fundamental unit carried by the electron. In this way, aggregates of both two and three quarks can make up a net charge of either zero or one. All quarks have spin ½, and so they are fermions. The masses of the quarks are not so well defined as those of other particles, because the energy of the glue which binds them into the hadron rivals the masses of the quarks themselves. Nevertheless, it is known that the strange quark is somewhat heavier than the up and down quark.

Trapped as they are inside hadrons, quarks can exist in excited states, much like the excited states of an atom but considerably more energetic. Indeed, the excess energy carried by an excited hadron contributes so much additional mass that before the quark theory was accepted physicists mistakenly believed they were dealing with completely distinct particles. Now it is accepted that many apparently different hadrons are actually only excited states of the same basic set of quarks.

As was explained in Chapter 5, quarks are firmly 'glued' together by the strong force. But they are also subject to the weak force. When the weak force acts on a quark it can change its flavour. This is the essence of the decay of a neutron. One of the down quarks in the neutron transmutes into an up quark, the excess charge being carried away by the electron which is simultaneously created. Similar flavour-changing, weak interactions bring about the decay of other hadrons.

The existence of the strange quark is needed to build a set of so-called 'strange' particles, heavy hadrons which were discovered in the early 1950s. The strange behaviour that prompted their name was due to the fact that these particles seem unable to decay using the strong force, even though both they and their products are hadrons. Physicists were baffled as to why, if all the particles remained in the hadron family, the strong force was inhibited from bringing about their decay. For some reason these hadrons had to resort to the much slower weak force. Why? The quark theory explained this puzzle naturally. The strong force cannot change the flavour of quarks; only the weak force can do that. Without a flavour change to convert the strange quark into an up or down, no decay can occur.

Table 2

Quark combination	*Name*	*Symbol*
uud	Proton	p
udd	Neutron	n
uds	Neutral sigma	Σ^0
dds	Negative sigma	Σ^-
uus	Positive sigma	Σ^+
uss	Neutral chi	Ξ^0
dss	Negative chi	Ξ^-
uds	Lambda	Λ
$u\bar{d}$	Positive pion	π^+
$d\bar{u}$	Negative pion	π^-
$d\bar{s}$	Neutral kaon	K^0
$u\bar{s}$	Positive kaon	K^+
$s\bar{u}$	Negative kaon	K^-
$s\bar{d}$	Neutral antikaon	$\bar{K}^0$

The three quark flavours up (u), down (d), and strange (s) possess charges +⅔, −⅓ and −⅓ respectively, and combine in threes to form eight baryons with the names shown. Quark–antiquark pairs combine to form mesons. (Some combinations, such as sss and $u\bar{u}$, have been omitted.)

Table 2 lists various possible quark combinations of three flavours and gives the names (usually restricted to a Greek letter) associated with each. The many excited states are not listed. The fact that various permutations of three basic particles could account for all the known hadrons was a major triumph for the theory. Despite this success, it was some years before direct physical evidence for the existence of quarks was forthcoming.

In 1969, in a series of historic experiments performed on the huge linear accelerator at Stanford in California (SLAC), this confirmation was obtained. The Stanford experimenters reasoned that if protons really have quarks inside them it might be possible to penetrate the proton and confront these particles face to face. What was needed was a subnuclear 'bullet' that could be fired right through the proton's interior. Another hadron would be useless, for it is as large as the proton itself. The ideal projectile is a lepton such as the electron. Because it does not feel the strong force, the electron will not get bogged down in the glue that binds the quarks together. On the other hand, it can still sense the presence of the quarks through their electric charge.

The Stanford experiment essentially consisted of using the 3-km accelerator as a giant electron microscope to build up an image of the proton's interior. In a conventional electron microscope, features less than a millionth of a centimetre can be discerned. A proton, however, is some ten million times smaller than this, and can only be probed by energizing the electrons to the equivalent of 2×10^{10} volts. At the time of the Stanford experiments few physicists subscribed to the simple quark theory. They certainly expected the electrons to be deflected by the protons' electric charges, but they assumed this charge was smeared out evenly within the proton. If that were so the pattern of electron scattering would be 'soft', i.e. the electrons would not be deflected especially violently during their passage through the protons. In the event, the scattering pattern was quite unlike this. It was as though some electrons were banging head-on into tiny rock-hard nuggets which sent them ricocheting off at wild angles. We now know that these solid lumps inside the protons are quarks.

In 1974 this simple version of the quark theory, which by that time was gaining ground among theorists, was dealt a sharp blow. Within a few days of each other two teams of American physicists, one at Stanford under Burton Richter and another at the Brookhaven National Laboratory under Samuel Ting, announced the independent discovery of a new hadron called psi. This in itself was hardly new, but the problem for the quark theory was that there was no room in the scheme for any more particles. Every possible combination of up, down, and strange quarks and their antiquarks was already accounted for. What was psi made of?

The problem was solved by appealing to an idea that had already been around for a while. There had to be a fourth quark flavour, which no one had seen until then. The new flavour had already been christened 'charm' (c), and the suggestion was that psi is a charmed meson, a charmed quark stuck to a charmed antiquark. Because antiquarks carry antiflavour, the charm quality is actually neutralized in the psi, and so a positive identification of charm had to await the production of other mesons where charmed quarks pair off with antiquarks of other flavours. Today a whole string of charmed particles are known. They are all very heavy, and so clearly the charmed quark is even heavier than the strange quark.

A re-run of this episode occurred in 1977 when the so-called upsilon came on the scene. This time there was little hesitation in invoking a fifth quark flavour, called 'bottom', or sometimes 'beauty' (b). The upsilon is a bottom quark–antiquark pair, so the 'bottomness' quality is hidden, but once again other quark combinations involving 'naked bottom' eventually turned up. Some idea of the comparative masses of the quarks can be gained from the fact that the pion is the lightest meson, made of up and down quark–antiquark pairs. The psi is some twenty-seven times heavier, while upsilon is no less than seventy-five times the mass of the pion.

The gradual growth in the list of known quark flavours parallels that of the various leptons, and a question that obviously arises is where it will all end. The quark theory was supposed to simplify our understanding of the hadron zoo, yet there is a distinct sense of proliferation again.

From the time of Democritus onwards, the underlying philosophy of atomism has been that, on a small enough scale, there exist truly elementary particles out of which all matter is built by aggregation. The attractiveness of this theory is that the fundamental particles, which are by definition indivisible, need come in only a handful of varieties. The complexity of matter is then explained as arising, not from the multiplicity of ingredients, but from the multiplicity of combinations. When it was discovered that dozens of different atomic nuclei exist, it shattered the hope that what we today call atoms were what the Greeks had in mind as elementary particles of matter. Though for historical reasons we still talk about the different chemical 'elements', atoms are now known not to be elementary at all, but composed of protons, neutrons, and electrons. If, now, the number of quarks grows too large there will be a temptation to suppose that they, too, must be composite bodies containing still smaller particles.

While there is some discontentment with the quark scheme for this reason, most physicists believe that the quarks are truly elementary particles, being point-like, indivisible, and with no internal structure. In this they resemble the leptons, and it has long been supposed that a deep connection must exist between these two distinct but similarly constituted families. The suggestions of a link can be discerned if the leptons and quarks are displayed as shown in Table 3. The leptons have been grouped together in pairs, each charged lepton going with its associated neutrino. The quarks have similarly been grouped into pairs. The table is arranged in such a way that each level represents a repetition of the one above. Thus, in level 2, for example, the muon is treated as a 'heavy electron', while the strange and charmed quarks are regarded as heavy versions of the down and up quarks respectively. A third level is also suggested, with the tauon a still heavier version of the electron and the bottom quark a really heavyweight version of the up quark. To complete the replication we still need another neutrino (the tauon-neutrino) and a sixth quark flavour, already dubbed 'top' or 'truth' (t). At the time of writing the experimental evidence for the top quark remains fragmentary, but few physicists doubt that it exists.

Table 3

Leptons	*Quarks*
e^-	u
ν_e	d
μ^-	c
ν_μ	s
τ^-	t
ν_τ	b

Leptons and quarks associate naturally in pairs according to their flavours, as shown. The permanent universe is made from the four particles in the top grouping. The two lower groupings seem to be merely duplicates of the first and (neutrinos excepted) consist of highly unstable particles.

Is it possible that there are fourth, fifth, . . . levels populated by still heavier particles? If so, the next generation of accelerating machines should produce them. There is a curious argument, however, which suggests that the three levels we already know are the only ones which exist. The argument concerns the number of neutrino types. We shall shortly see that during the hot big bang which marked the origin of the universe, neutrinos were produced in great abundance. A sort of principle of democracy ensured that equal shares of energy went to each separate species of particle, and so the greater the number of neutrino types the more the energy that became locked up in the sea of neutrinos which permeates the cosmos. Calculations indicate that the gravity of all these neutrinos would have had a serious, perturbing effect on the nuclear processes which occurred in the first few minutes of the primeval universe, if more than three neutrino species exist. Therefore, by this indirect argument, it seems likely that the three levels of structure shown in Table 3 represent the totality of hadrons and leptons that nature employs.

It is a curious fact that all ordinary matter in the universe is made from just the two lightest leptons (the electron and its neutrino) and the two lightest quarks (up and down). If the other leptons and quarks suddenly ceased to exist it is probable that very little would change in the world.

The remaining quarks and leptons seem to be simply unnecessary duplicates of this top level of structure. They are all unstable and rapidly decay into top-level particles. Thus, the tauon and muon decay into electrons, while the strange, charmed and bottomed particles soon disintegrate, either into neutrons or protons in the case of the baryons, or into leptons in the case of mesons. This prompts the question, what are all these particles in levels two and three *for*? Why does nature bother with them?

The messenger particles

The six pairs of leptons and quarks, while accounting for the building blocks of matter, do not exhaust the list of all particles known to physicists. Some particles, such as the photon, are not included in this scheme. These remaining particles are not part of the building material of the world, but are related to the 'glue' that holds the world together, i.e. they are associated with the four forces.

I remember being told when I was a boy that the moon made the oceans rise and fall in the daily tides. It always seemed mysterious to me that the water in the ocean should *know* where the moon is, and thus follow its motion across the sky. When as a student I learned about gravity the sense of bafflement only deepened. How did the moon manage to reach out across a quarter million miles of empty space and grasp hold of the ocean? The standard answer – that the moon produces a gravitational field in its vicinity and this field touches the ocean and persuades it to move – made some sense. But still I was not satisfied. You can't see the moon's gravitational field. Wasn't it just a fancy way of talking? Did it really explain anything? It seemed to me that somehow the moon had to tell the ocean it was there. Some sort of message or signal had to pass between them so that the water would know how to move.

As it turned out, the idea of a force being communicated across space in the form of a signal is not so far from the modern approach to the subject. To understand how this picture comes about it is necessary to look at the mechanism of a force field in greater detail. We shall take as an example, not the ocean tides, but the simpler case of two electrons that approach each other, experience an electric repulsive force, and accelerate apart. This is known to physicists as a 'scattering problem'. Of course, the electrons do not literally bounce off each other; they don't actually touch. All the forcing is done at a distance through the electromagnetic field which emanates from each electron.

It is easy to build up a mental picture of electron scattering. Initially, the particles are far apart and their effect on each other is weak. They move on converging paths that are almost straight (Figure 11). Then, as the re-

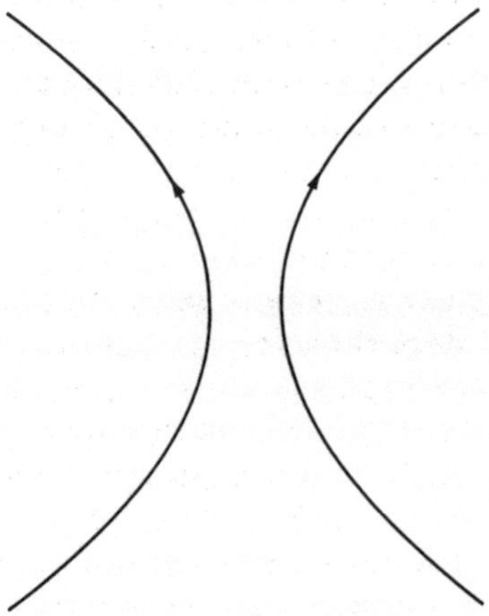

Figure 11. Scattering of two charged particles is depicted by showing their paths curving away from each other on close approach, due to the force of electric repulsion.

pulsive force builds up, the paths begin to curve until the electrons reach a point of closest approach, after which the paths bend away from each other and the electrons recede, eventually returning to straight line motion but along diverging trajectories. This sort of scenario could easily be demonstrated in the laboratory using electrostatically charged balls in place of electrons. Again we have the problem of how each particle 'knows' the other one is there so that it can adjust its motion in response.

Although the picture of the curving electron paths is easy to visualize, it is badly wrong on a number of counts. This is because electrons are quantum particles, and their behaviour is subject to all the peculiarities of quantum physics. For a start, electrons do not follow well-defined paths in space anyway. We can perhaps determine their points of departure and arrival, before and after the scattering event, but the tracks followed in between are obscure and indeterminate. Furthermore, the intuitive idea of the electron exchanging energy and momentum with the field in a continuous fashion as it is accelerated conflicts with the existence of photons. Energy and momentum can only be transmitted through the field in packets or quanta. A more accurate picture of the way in which an electron's motion is disturbed by the field is to suppose that the particle experiences a sudden jerk by absorbing a photon from the field. Viewed at the quantum level, then, the scattering event between two electrons can be depicted as in Figure 12. The wavy line joining the two electrons' paths represents a single photon emitted by one electron and absorbed by the other. The scattering event is now shown as an abrupt change in the direction of each electron's motion.

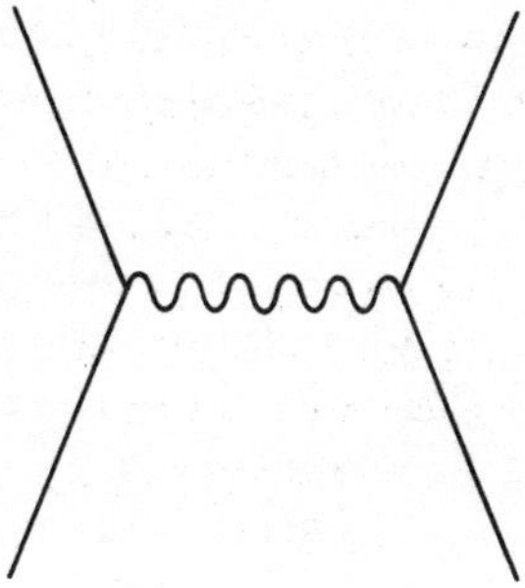

Figure 12. Quantum description of charged-particle scattering depicts the force as conveyed by a messenger, or 'virtual', photon (wavy line) exchanged between the particles.

This sort of diagram was first used by Richard Feynman to represent terms in an equation, and was intended to have symbolic value only. Since then, Feynman diagrams have come to be used informally as a crude picture of what is actually supposed to be happening physically. Such pictures are a great aid to intuition but should be interpreted only loosely. For example, it is never possible to observe the sharp kink in the path of an electron. If we observe only its ingoing and outgoing positions we do not know the exact moment when the photon is exchanged, nor do we even know which particle does the emitting and which the absorbing. All these details are lost in the mists of quantum fuzziness.

In spite of this caution, Feynman diagrams give a powerful representation of the quantum version of a force. We can think of the photon that is exchanged between the electrons as a sort of messenger particle sent out from one electron to tell the other, 'I'm here, so move!' Of course, all quantum processes are subject to the rules of probability, and so such an exchange will only occur with certain definite odds. It may be that the electrons exchange two or more photons (Figure 13), though this is less likely.

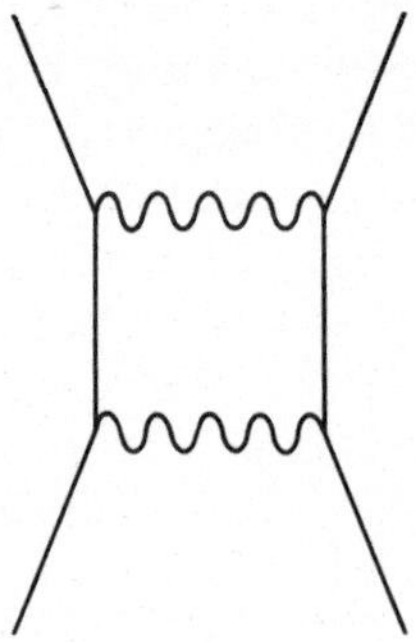

Figure 13. Two electrons scatter by the exchange of two messenger photons. Such processes represent only a small corrective force to the dominant process depicted in Figure 12.

It is important to realize that we do not actually see the messenger photons passing back and forth from one electron to another. The messengers remain a sort of private arrangement between the electrons. They exist solely to tell the electrons how to move, and though they carry energy and momentum, the usual rules for these quantities familiar from classical physics need not be obeyed. Messenger photons have been compared with the ball that is exchanged between tennis players. Just as a tennis ball shapes the pattern of activity followed by the players, so the photon influences the behaviour of the electrons.

The success of the messenger particle description involves widening the concept of photon, from a particle of light that we can see, to a rather ephemeral entity 'seen' only by the charged particles being scattered. Sometimes the photons that we see are called *real* and the messenger photons *virtual* to remind us of their temporary, almost ghostly quality. The distinction is actually somewhat artificial, but it is in widespread use.

The description of electromagnetic activity in terms of virtual messenger photons amounts to far more than a picturesque model of quantum forces. It is actually a highly sophisticated and detailed mathematical theory known as *quantum electrodynamics*, or QED for short. When

QED was first formulated properly, shortly after the Second World War, physicists had available a theory which obeyed the principles of both the quantum theory and the theory of relativity. Here was an excellent chance to see both these important aspects of the new physics working together, and to test them in practical experiments.

Theoretically, QED was a remarkable accomplishment. Earlier work on the interaction of photons and electrons had met with only limited success due to mathematical difficulties. Once the theorists eventually learned how to do their sums correctly everything fell beautifully into place. The theory offered a procedure for calculating the results of any desired process involving photons and electrons, however complicated.

To test how well their new theory matched the real world, physicists focussed their attention on two particularly interesting physical effects. The first concerned the energy levels of the hydrogen atom, the simplest atomic system. According to QED, the levels ought to be very slightly shifted from where they would be if virtual photons did not exist. The theory yielded a very precise value for the predicted shift. An experiment to look for the shift and to measure its value as accurately as possible was performed by Willis Lamb of the University of Arizona. To universal delight, calculation and experiment matched exactly.

The second decisive test of QED involved a very small correction to the magnetic field carried by the electron. Once again theory and experiment were in complete accord. Theorists set about refining their calculations, and experimenters improved their techniques. As the accuracy of both got better and better, the agreement remained flawless. Today experiment and theory agree to the limits of their precision, which is better than nine-figure accuracy. This astonishing consistency makes QED the most successful quantitative scientific theory in existence.

In view of these triumphs it is easy to understand how QED became the model for the quantum description of the other three forces of nature too. Of course, the different fields associated with the other forces required different sorts of messenger particles. In the case of gravity, a particle called the *graviton* was invented, which plays a role analogous to the photon. When two particles exert a gravitational influence on each other, gravitons are exchanged between them. We can draw diagrams very similar to Figures 12 and 13 to represent this. It is the gravitons which carry the moon's message to the oceans telling them to rise and fall in tidal motion. Gravitons speeding back and forth between the Earth and the sun keep our planet in its orbit. A network of gravitons binds you and me firmly to the Earth.

Like photons, gravitons travel at the speed of light; they are therefore 'zero rest-mass' particles. But here the resemblance ends. Whereas photons carry one unit of spin, gravitons carry two. This is an important difference because it determines the direction of the force. In electromagnetism identical particles such as electrons repel each other; for gravity, all particles attract.

Gravitons also come in real and virtual varieties. A real graviton is a quantum of a gravitational wave, just as a real photon is a quantum of an electromagnetic wave. In principle we could 'see' real gravitons. However, because gravity is such an incredibly weak force, gravitons are impossible to detect directly. They couple to other quantum particles with a strength which is so low that the probability of a graviton being scattered or absorbed, say by a proton, is infinitesimal.

Table 4

Force	*Name of messenger*	*Charge*	*Mass*
Electromagnetism	Photon	0	0
Gravity	Graviton	0	0
Weak	$W^{\pm}$	± 1	85
	Z	0	95
Strong	Gluon	0	0

The messenger particles which convey the four forces of nature. Mass is expressed in units of the proton mass.

Turning to the remaining forces (see Table 4), weak and strong, the basic idea of messenger particle exchange can be maintained. However, there are important differences of detail. The strong force, it will be recalled, is responsible for gluing the quarks together. This can be achieved by a force field similar to electromagnetism, but more complicated. Electrical forces can cause two particles with unlike charges to bind together. In the quark case, bound states form containing three particles, which suggests a more elaborate sort of force field involving three varieties of 'charge'. The messenger particles which commute between the quarks, sticking them together in pairs or trios, are called *gluons*.

In the case of the weak force the situation is a little different. The force is extremely short in range. To achieve this, the weak force uses messenger particles which have a large rest mass. The energy locked up in this mass has to be 'borrowed' using the Heisenberg energy uncertainty principle already discussed on page 42. Because the mass (hence energy) borrowed is so great, the rules of the uncertainty principle require the loan to be very short, lasting only about 10^{-26} s. Being so short-lived, the messenger particles don't go very far, and the force is therefore very short in range.

There are actually two different species of weak messengers. One is identical to the photon in everything except rest mass. It does not have a name, but is known only by the letter Z. The Z particle is essentially a new form of light. The other type of weak messenger particles are called W and differ from Z in that they carry electric charge. In Chapter 7 the Z and W, which were only discovered in 1983, will be discussed in more detail.

The classification of particles into quarks, leptons, and messengers completes the list of known subatomic particles. Each plays a separate yet crucial role in shaping the universe. Without the messengers there would be no forces and every particle would remain oblivious of its neighbours. No structures could exist and no activity of any consequence would occur. Without quarks there would be no atomic nuclei and no sunlight. Without leptons atoms could not exist, and there would be no chemistry and no life.

What is the point of particle physics?

The prestigious British national newspaper the *Guardian* recently carried an editorial questioning the motivation behind particle physics, a multi-million dollar enterprise that soaks up not merely a sizeable fraction of national science budgets but also a fair proportion of intellectual talent. Did the physicists know what they were doing, asked the *Guardian*. If they did, what use was it anyway? Who but the physicists care about all these particles?

Coincidently, a couple of months later I was attending a lecture in Baltimore given by George Keyworth, science adviser to the U.S. President. Keyworth also addressed the subject of particle physics, but the tone was very different. American physicists were chagrined about a recent announcement from Europe's showpiece particle laboratory, CERN, that the all-important W and Z particles had at last been discovered on their large proton-antiproton collider. Americans had grown used to their own high-energy labs scooping all the new discoveries. Was this a sign of scientific, even national decline?

Keyworth was in no doubt that the health of America generally, and her economy in particular, demanded that the nation be at the forefront of scientific endeavour. Major projects in fundamental physics lay at the cutting edge of that endeavour, he said. America had to regain her particle physics supremacy.

That same week the news wires buzzed with the story about a monster American accelerator being planned for a new generation of particle physics experiments. The basic cost was put at $2 000 million making it by far the most expensive single machine ever built by man. Dwarfing even CERN's giant new accelerator LEP, Uncle Sam's behemoth could be large enough to encircle the whole of Luxembourg! Huge superconductors would generate the intense magnetic fields needed to deflect the particle beams round the ring-shaped tube, which would be so obtrusive it would have to be located deep in the desert. I wondered what the editor of the *Guardian* thought.

Known as the Superconducting Super Collider, but popularly dubbed 'the Desertron', this awesome machine would be capable of accelerating protons to twenty thousand times their rest-mass energy. There are many ways of interpreting this statistic. At top speed the particles will fall short of the speed of light – the ultimate speed in the universe – by only 1 kilometre per hour. The effects of relativity will be so great that each particle will weigh twenty thousand times more when in flight than at rest. From the viewpoint of such a particle, time will be stretched so much that 1 s to the particle will correspond to 5.5 h in our frame of reference. Each kilometre of tube round which it plunges will appear to the particle to be shrunk to a mere 5.0 cm.

What is the compelling urge for nations to pour such vast resources into smashing up matter ever more violently? Do these investigations have any conceivable practical use?

Doubtless there is a strong nationalistic flavour involved in all big science. As in the arts or sport, it's nice to win the prizes, and world acclaim. Particle physics has become something of a national virility symbol. If you do it well, and in spectacular style, it shows that there is nothing much wrong with your science and engineering, or your economy. That bolsters confidence in your other technological products of a more exportable variety. The actual construction of an accelerator and all its attendant gadgetry demands a very high level of technological competence. Valuable experience is gained in new technologies which can have important spin-off in other scientific enterprises. For example, the research and development work for the superconducting magnets needed for the Desertron has been in progress in the U.S. for twenty years. Nevertheless, these are all indirect benefits, hard to measure. Are there any tangible results?

Another argument one sometimes hears in support of fundamental research goes something like this. Physics tends to run about fifty years ahead of technology. When new discoveries are made, possible applications are often far from obvious, yet few major advances in our understanding of basic physics have failed to be exploited sooner or later. Think of Maxwell's work on electromagnetism: could he have foreseen modern telecommunications and electronics? What about Rutherford's statement that he did not think nuclear energy would ever have a use? Who can tell what might come out of the study of particle physics, what new forces might be released, what new principles discovered that would extend our understanding of the world and give us power over a wider range of physical conditions? The applications could be as revolutionary as radio or nuclear power.

Most branches of science are eventually exploited at some stage as military technology. So far, particle physics (in contrast to nuclear physics) has remained relatively innocent in this respect. Keyworth's address, however, happened to coincide with a wave of publicity about President Reagan's controversial anti-missile project, colloquially known as the beam weapon programme. The proposal is to harness high-energy particle beams so that they can be directed against enemy missiles. Here, then, would be a practical, if sinister, application of particle physics.

The prevailing opinion is that such a device is unfeasible (although rumours persist that the Russians are already working on one). Certainly most particle physicists find the idea both absurd and abhorrent, and they have reacted strongly against the President's proposal. Keyworth castigated them for this, and appealed to each member of the physics community 'to consider what part you can play' in making beam weapons a working proposition. His entreaty (no doubt coincidentally) followed his remarks on the funding of high-energy particle physics.

It has always been my opinion that physicists should not feel compelled to defend basic research by appeal to spin-off (most especially of the military variety), historical analogy, or vague promises about miracle gadgets. Physicists conduct their research primarily for its own sake, from a deep sense of curiosity about how the world is put together, and for a desire to know and understand nature in ever greater detail. Particle physics is a human adventure story without parallel. For two and a half millenia mankind has sought the ultimate building blocks of matter, and now we are close to the end of that goal. With colossal machines we can peer into the very heart of matter, and wrest from nature her innermost secrets. There may be spin-off, there could be undreamt-of technology;

7
Taming the Infinite

The road to unity

A visit to a large particle accelerator laboratory is always a stimulating experience. The hundreds of scientists, engineers, and support staff, the concentration of talent dedicated to the operation of a small number of huge machines awesome in their power and complexity; the sheer size and commitment of the enterprise; the sense of probing the unknown – all combine to give the impression of science at its most refined.

When I visited CERN to give some lectures in the autumn of 1982 it was clear that something special was going on. There was an atmosphere of excitement and expectancy, a feeling that major discoveries lay just around the corner. Superficially it was business as usual: bustling corridors, scientists scurrying to experimental areas or seminars, overseas visitors engaged in urgent conversation, gleaning the latest information about this and that, theorists busily scrawling equations on blackboards or endless sheets of scrap paper, secretaries at their typewriters, technicians peering at wires and tubes, the clatter and chatter of the lunchtime cafeteria. At any one time, many experiments are going on at CERN The accelerators themselves, housed in bland concrete-lined buildings or invisible in their subterranean cocoons, give no outward sign of activity. The casual passer-by might suspect nothing of the forces being unleashed inside the bowels of these technological titans.

The biggest of CERN's machines at that time was the proton – antiproton collider, a ring-shaped tube several miles in circumference, in which protons and antiprotons circulate in opposite directions. When the particle beams have been boosted to the right energy, they are brought into head-on collision. Protons then annihilate antiprotons, their energies being released to produce showers of new particles that spew out in all directions from the point of impact. The particles' tracks are often revealed by special electronic devices which are triggered by the passage of electrically charged bodies. By using an array of such devices a three-dimensional reconstruction of the impact event can be obtained. The machine is fed protons and antiprotons from a battery of peripheral devices, one of which makes the highly explosive antiprotons and stores them in a special magnetic ring before injecting them into the collider. The system was masterminded by an Italian physicist, Carlo Rubbia. In the autumn of 1982 Rubbia was the man of the moment.

Although the collider was then still being eased up to its design capability, impatient physicists were eagerly trying to glimpse a preview of the sort of events collisions at these unexplored energies would create. Despite moving into uncharted territory, the CERN team still had theory as a guide, and one anticipated landmark towered high above all the others. If the sums were right, at any day the physicists were about to catch their first sight of a new type of particle – the W – messenger of the weak force. The W had been predicted decades before but never seen. A confirmation of its existence would mark the first step on the road to the superforce.

By December the experimenters were sure. Rumours buzzed around the world. And then, in the middle of January 1983, Rubbia called in the press. The W had been found.

In its essence, all science is a search for unity. The scientific method owes its remarkable success to the scientist's knack of linking together fragments of knowledge into a pattern. Forging links is one of the most fundamental, and satisfying, of scientific pursuits. The link between gravity and the motion of the planets found by Newton heralded the very birth of the scientific era. The connection between microbes and disease established modern medicine as a real science. The link between the thermodynamic properties of a gas and the chaotic agitation of its molecules established the atomic theory of matter. The relationship between mass and energy opened the way to nuclear power.

Every time new links are forged, our understanding of the natural world widens and our control over it is extended. New links do more than just unify a body of knowledge. They open up pathways to hitherto unsuspected phenomena. Linkage is therefore both a synthesis of knowledge and a stimulus, propelling scientific endeavour towards rich, new pastures.

Fundamental physics has always led the way in unifying knowledge. But what has happened since the early 1970s is beyond comparison. We seem to be on the threshold of a unification more powerful and more profound than anything that has gone before. There is a growing belief among physicists that we are beginning to glimpse nothing less than a unified theory of all existence.

Complete theories of the universe are not new. Most religions purport to describe the natural and supernatural worlds in their cosmic totality. But mystical cosmologies are rooted in ancient wisdom, divine revelation and complex theology. No two of them are the same.

Scientific theories of this sort are rare, though not unknown. The British astronomer Arthur Eddington, for example, attempted to build an all-encompassing description of matter, force, and creation in his book *Fundamental Theory* published in 1946. Eddington's ambitious ideas were, however, very much the personal dream of an isolated and perhaps somewhat eccentric scientist. Now, for the first time, both scientific experiment and theory have advanced to the point where a complete theory of the universe can be given that rests on testable and widely accepted hypotheses.

The underlying stimulus for this great leap forward comes from a study of the fundamental forces of nature. In Chapter 5 it was explained how physicists recognize four fundamental forces: gravity, electromagnetism, the weak force, and the strong nuclear (or gluon) force. Already in the 1850s Maxwell had united electricity and magnetism into a single electromagnetic theory. Then in the 1920s Einstein began a systematic attempt to unify electromagnetism with his new theory of gravity (the general theory of relativity).

Before long he was overtaken by events. The nuclear forces, weak and strong, were discovered, and any attempt to unify the forces of nature therefore had to contend with four rather than two fundamental forces. Yet the fascination remained. Why should there be four distinct forces? The prospect of describing all natural activity in terms of a single, unified superforce remained a compelling, if somewhat remote dream. Today that dream is no longer so remote. It could soon become a reality.

The decisive step on the modern road to unity was taken in the late 1960s. By that time theorists had enjoyed unprecedented success in applying the quantum theory to fields. The field concept had been invented over a century before, and had already proved itself in a wide range of practical applications, such as radio-engineering. The marriage of quantum mechanics and the electromagnetic field led directly to quantum electrodynamics (QED) with its legendary accuracy and predictive power.

Unfortunately, the same success could not be claimed for the other three forces of nature. The quantum theory of gravity, in which graviton messengers transmit the gravitational force, was bogged down in mathematical complexities. The weak force remained poorly understood. The existence of Z had not been accepted, while a description in terms of W particle exchange alone gave sensible results for only the simplest types of low-energy processes. The strong force was still less well understood. It had become clear by then that hadrons in general, and protons and neutrons in particular, were not after all elementary particles, but the quark theory was not then firmly established. The forces between hadrons were clearly very complicated, but nobody knew how to model the internal structure of hadrons to find a simpler underlying description.

So it was that in the 1960s each of the four forces of nature was described by a different type of theory, and of these only one, QED, could in any sense be described as successful. Theorists began to wonder where the secret of QED's success lay. What were the features enjoyed by the electromagnetic field, but not the other force fields, which enabled a quantum description to work so well? If these features could be identified it might be possible to modify the theory of the other forces to incorporate the crucial ingredients.

The living vacuum

Empty space does not appear a very promising subject for study, yet it holds the key to a full understanding of the forces of nature. The idea of a vacuum is easy enough to visualize: it is a region of space from which everything has been removed, all particles, all fields, all waves. In practical terms a perfect vacuum is impossible to achieve. Even in outer space there is always a tiny residue of gas or plasma, as well as the universal background radiation left over from the big bang. In spite of this we can still discuss a vacuum as an idealization.

When physicists began to study the quantum theory of fields, they discovered that a vacuum was not at all what it had long appeared to be – just empty space devoid of substance and activity. Quantum physics seemed capable of playing tricks even in the absence of any quantum particles.

The source of the trickery can be traced to Heisenberg's uncertainty principle as it relates to the behaviour of energy. In Chapter 2 it was explained how the law of energy conservation can be suspended by quantum effects for a very short interval of time. During this brief duration energy can be 'borrowed' for all manner of purposes, one of which is to create particles. Any particles produced in this way will be pretty short-lived, because the energy tied up in them has to be repaid after a minute fraction of a second. Nevertheless, particles are permitted to pop out of nowhere, enjoying a fleeting existence, before fading once again into oblivion. This evanescent activity cannot be prevented. Though space can be made as empty as it can possibly be, there will always be a host of these temporary particles whose visit is financed by the Heisenberg loan. The temporary 'ghost' particles cannot be seen, even though they may leave physical traces of their brief existence. They are, in fact, a form of 'virtual' particle, similar to messenger particles, but with nothing on the 'ends of the line' to send or receive the message. They travel from emptiness to emptiness, witnesses to the existence of a force field, but with nothing permanent to push against.

What might appear to be empty space is, therefore, a seething ferment of virtual particles. A vacuum is not inert and featureless, but alive with throbbing energy and vitality. A 'real' particle such as an electron must always be viewed against this background of frenetic activity. When an electron moves through space, it is actually swimming in a sea of ghost particles of all varieties – virtual leptons, quarks, and messengers, entangled in a complex mêlée. The presence of the electron will distort this irreducible vacuum activity, and the distortion in turn reacts back on the electron. Even at rest, an electron is not at rest: it is being continually assaulted by all manner of other particles from the vacuum.

If two electrons exchange a messenger photon the transaction is but a further disturbance amid a pre-existing powerhouse of interchanges. A proper description of the forces between particles must take into account all these additional virtual quanta. The total experience of a given particle in the presence of force fields will include processes where two, three, or more messengers are exchanged, where messengers interact with vacuum particles, and where vacuum particles cling to the transmitting and receiving particles. There will be an infinte number of possible interactions all going on at once.

The process depicted in Figure 14 is a relatively simple example of one of these higher-order processes. One of the source particles emits a virtual photon, which then produces an electron–positron pair. The members of this pair then exchange another virtual photon between them, before annihilating each other to form another virtual photon which is in turn absorbed by the receiving particle. It could be that this whole diagram is itself only a fragment of a still more elaborate picture, in which the two source particles exist only temporarily before changing into something else.

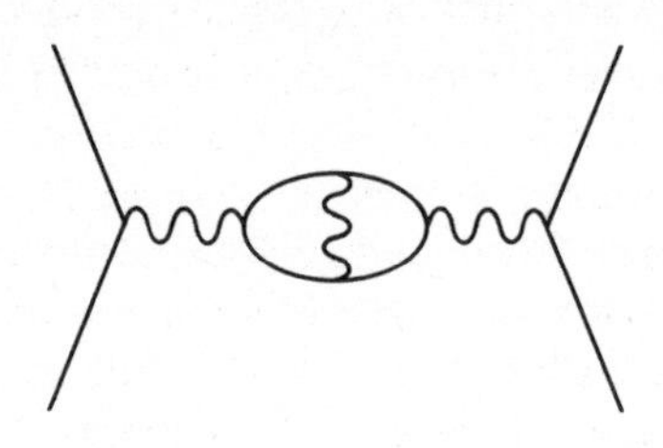

Figure 14. Complex interaction between two particles due to the exchange of a messenger photon which itself interacts with other messenger particles on the way.

We must envisage the interactions between all particles as produced by webs of increasing complexity, made up of ever more convoluted exchanges between different sorts of virtual particles. A force field is, in fact, never static. There are always ghost particles coming and going, appearing and disappearing, entwined in a shimmering pattern of energy.

At first it seems as if the infinite complexity of all this activity must prevent any hope of our understanding, let alone computing, the forces between real particles. Fortunately, this is not the case. It turns out, at least in QED, that as the processes become more and more complex, so their effect on the real particles diminishes. In the example of the scattering of two electrons the dominant contribution comes from the exchange of a single messenger photon. The other processes just lead to small corrections. In a practical calculation it is unusual to consider more than three or four of the simplest diagrams unless very high accuracy is required.

Imagine dropping a new particle into this restless sea of vacuum activity. Immediately, the additional particle envelopes itself in a quivering shroud of energy. We cannot actually see this cloak of energy directly, but let us make-believe that we have a magic microscope, capable of detecting all the virtual quanta. Looking through the microscope we see at the centre the 'bare' particle, which we shall suppose to be an electron. On the outer fringes of the cloud which surrounds it low-energy photons weave about, probing the space around the electron and tangling with the ghostly half-forms of the vacuum, merging into the shifting sea of virtual quanta that pervades all of space.

As we probe deeper into the shroud, so the virtual photons are seen to be more energetic, their activities more agitated. Some of the photons occasionally transmute into pairs of electrons and positrons, which then rapidly re-fuse to give back a photon. Sometimes a more complicated exchange takes place, involving still more virtual particles. Closer still to the electron the shroud is found to be positively pulsating with energy. Here, heavier particles mingle incoherently with the photons; we can see quarks, heavy leptons, messenger particles of all varieties.

Turning up the power of the microscope we notice that the energy in the shroud rises rapidly as the electron itself is approached, rising seemingly without limit. There seems to be a great crisis.

The miracle of QED

It is sometimes said that out of every crisis in science a new theory is born. The crisis revealed by our imaginary microscope is a symptom of a deep-rooted malaise in the field concept. In spite of its impressive successes the quantum theory of fields comes badly unstuck at one crucial point.

The origin of the difficulty can actually be traced back to classical physics. A planet such as the Earth produces a gravitational field which, for example, acts on the moon. In addition to this, the Earth's field also acts on the Earth itself. The material beneath our feet is gripped by the Earth's own gravity. Such 'self-action' seems to be an unavoidable consequence of field theory.

It was once believed that a particle such as an electron was a scaled-down version of the Earth, a tiny solid ball with electric charge spread evenly throughout it. Just as the Earth acts on itself through gravity, so an electron must act on itself through electric forces. The two cases differ in one important respect, however: gravity is attractive and holds the Earth together; the electric forces within the electron are repulsive, and try to blow the electron to pieces. In view of this, there arose the vexing problem of what other forces could be present to counteract the electric repulsion and so retain the electron in one piece.

With the arrival of the theory of relativity the problems of modelling the electron in terms of a tiny solid ball were compounded, for there was an added difficulty concerning the assumption that the electron is a rigid body. Imagine striking a spherical ball a sharp blow so that it flies off in some direction. If the ball were perfectly rigid it would move without change of shape. To achieve this, all regions of the ball have to begin moving simultaneously. But this contradicts the principle that no physical influence can travel faster than light. The region of the ball remote from the point of impact cannot know about the blow, and so cannot respond, until at least the time taken for the shock wave to travel – no faster than light – across the interior of the ball. Conversely, the region of the ball near the point of impact must start to move before the remainder of the ball. The ball will therefore change shape; it cannot be perfectly rigid. But if the electron can be squeezed and squashed it could in principle be pulled to pieces. It would not be an elementary particle at all. We should expect to see bits and pieces of charged matter of all shapes and sizes, whereas in fact electrons are indistinguishable.

To escape from this dilemma physicists were compelled to abandon the idea of an electron as a solid ball. Instead, they came to regard it as a structureless point with no extension at all. Although this alleviated the problem of how the electron's internal parts hold together, it brought with it a further source of difficulty. This time the trouble arose from the field surrounding the electron. The electric force of a charged body diminishes with distance in accordance with the inverse square law. Conversely, the field builds up in strength close to the source. In the case of a point source the field strength rises without limit as the source is approached. This means that the total electrical energy of the system is infinite.

The existence of infinite field energy associated with a point electron seems to deal a death blow to field theory. If the electron possessed infinite energy it would be infinitely heavy, which is absurd. Theorists were faced with a choice: they could either abandon the point electron model or find a way around the impasse. A way around was found, though to some people it seemed like cheating. It is known as 'renormalization'.

Imagine that by some magic we could 'switch off' the charge on an electron. The field emanating from it would disappear, the electrical energy going with it. We would be left with what could be termed a 'bare' electron, stripped of the electromagnetic field that clothes it. One can ask what the mass of this bare electron is. The observed mass of a real electron will be composed of two pieces: its bare mass plus the mass of the electric energy generated by the field. The embarrassment is that the electric energy part is calculated to be infinite. This would clearly be nonsensical if we could actually switch off the electric charge of the electron since no physical quantity can change by an infinite amount. However, we cannot do this. The charge can't be switched off. When we observe an electron we perceive the whole package, field and all. And the *observed* mass is, of course, finite. So do we really have to worry if our sums tell us that an inseparable part of the electron's mass is arithmetically infinite?

The answer is that some people do worry, but not too seriously. The presence of infinite terms in the theory is a warning flag that something is wrong, but if the infinities never show up in an observable quantity we can just ignore them and go ahead and compute. To do this the infinities have to be squashed out of the calculation somehow in order that the computation can proceed. In practice the theorist simply adjusts, or 'renormalizes', the zero-point on the scale used for measuring mass by an infinite amount. It's rather like agreeing to measure the height of an aeroplane from ground level rather than sea level, only in the case of an electron the shift involved is an infinite quantity. The theorist explains that it doesn't matter since there is no god-given zero point on the mass scale anyway; any change in our choice – even an infinite change – will never be observable in the real, physical world.

By this mathematical sleight-of-hand the description of the electron can be rid of the infinite terms that at first threatened to reduce the theory to absurdity. Further trouble was in store, however, when a quantum description of a point-like electron was attempted. The problem now centres on the nature of virtual photons.

We have seen how every electron is clothed in a shroud of shimmering quantum energy made up of all sorts of virtual particles. Let us take a closer look at the cause of this shroud. Originally, virtual photons were introduced as a quantum description of how one electron signals another that it wishes to exert a force. It is possible, however, for a solitary electron to *act on itself* with virtual photons. In the classical theory such self-action also exists and produces infinite terms if the electron is treated as point-like. A quantum description of self-action means that, picturesquely speaking, an electron sends a messenger photon to itself. A diagram depicting this self-action is shown in Figure 15. It shows a virtual photon emitted by an electron, venturing a short way into space, and then coming back to be re-absorbed by the same electron. The idea of a photon turning round like this comes as a surprise, but it must be remembered that common sense notions do not hold in the quantum realm where rebellion is the norm.

In the quantum description of an electron, therefore, the electromagnetic field which clothes the particle must be viewed as a retinue of virtual photons fussing around the electron, clinging to it, and forming a tenacious shroud of energy. The photons come and go very rapidly. Those that remain close to the electron, near the centre of the shroud, carry considerable energy; in fact, when the total energy of the photon shroud is computed, it again turns out to be infinite.

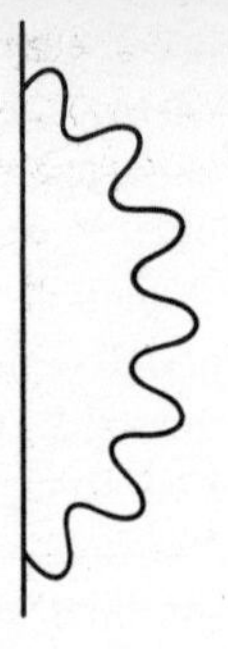

Figure 15. A charged particle emits and then re-absorbs its own messenger particle. Such processes lead to 'self-action' which endows the charged particle with energy. Mathematically the total quantity of energy produced by all such loops is infinite.

Presented with this result, the theorist can proceed to 'renormalize' the infinite energy away, just as he did in the classical theory. This time, however, things are not so simple. The loop shown in Figure 15 is actually

Figure 16. More complicated self-action processes lead to further infinite energy effects. In QED all such effects, however complicated, can be side-stepped mathematically by a single subtraction operation ('re-normalization').

only one possible process whereby an electron can act on itself. More complicated self-action loops are conceivable, as for example the one shown in Figure 16. Here the messenger photon creates a virtual electron–positron pair *en route* round the loop. It is obvious that as more and more complicated loops are entertained, there is no end to the total number of ways an electron can act on itself with messenger particles. Now each of these loops contributes its own infinite amount of energy to the system. Every conceivable web of loops produces infinite energy. Rather than contending with just a single infinity, as in the classical theory, there is now an unending sequence of infinite terms in the calculation. At each stage we could attempt to knock out an infinity by artificially subtracting an infinite term, but as soon as we have done so another infinity pops up again. There seems to be no escape.

It is in the face of this bleak prospect that something of a miracle occurs. When this horrendous sequence of infinite terms is packaged up the right way (mathematically speaking) it so happens that the whole lot can be disposed of in one go. A single infinite subtraction, or renormalization, wipes every infinity away no matter how complicated the loop that produced it. Of course, demonstrating that this miracle works was something of a *tour de force* when it was worked out thirty years ago. Without it, the theory would crumble into meaninglesss nonsense.

Naturally the theorists were delighted with this result. It is good to know that something really works, that there is no longer anything mysterious about the way in which electrons and photons interact. They called QED a 'renormalizable theory' and went on to confirm those tiny effects due to the virtual particles which were so accurately checked by experiment, such as the shift in the energy level of the hydrogen atom measured by Lamb, and the small correction to the electron's magnetic field. The extraordinary agreement between theory and experiment at this level of detail showed that all those virtual particles and vacuum effects are not just a theorist's invention, a figment of imagination. They are really needed to achieve an accurate description of the atomic world.

Spurred on by this remarkable success, the theorists turned their attention to the other forces of nature to see if the renormalization trick would work for them too. Each force field generates its own set of infinite energies (and other infinite quantities besides). The hope was that the miracle of the vanishing infinites found in QED would work for these forces too.

Sadly, that hope was misplaced. Of the four forces of nature, only electromagnetism seemed to have the magic property of renormalizability. The messenger particles of the other forces – as they were conceived at the time – generated endless infinities that could not be swept away *en masse* as in Q.E.D. The theorists went back to the drawing board to try and understand the secret of QED's success. It soon became apparent that it all had something to do with symmetry.

Symmetry points the way

The historian and writer C. P. Snow has written of the 'two cultures' into which modern technological society divides – roughly, the scientific and the artistic. Many scientists, however, have sensitive artistic taste. They enjoy paintings and sculptures, they frequently play musical instruments with unusual ability, and they tend to have a deep appreciation of style and beauty. For theorists especially, science itself can become an art form, a subtle blend of mathematics and miracle.

In Chapter 4 we saw how aesthetic judgement plays such an important part in scientific advance. Among the more successful examples of this is the application of symmetry, in a sufficiently general sense, to fundamental physics. Indeed, in recent years the symmetry bonanza has proved so powerful that it has taken over the thinking in whole areas of the subject. It now seems certain that symmetry is the key to a proper understanding of the forces of nature. Physicists now believe that all forces exist simply to enable nature to maintain a set of abstract symmetries in the world.

What has a force got to do with symmetry? The proposition seems baffling and obscure. A force is something which pushes and pulls on matter, or transmutes the identities of particles. Symmetry is a quite different concept, related to harmony and simplicity of form.

To find the answer to this question let us first recap on what is meant by symmetry. In broadest terms something is symmetric if it remains unchanged under a certain operation. A sphere is symmetric because it looks the same when it has been rotated by any angle about its centre. A cathedral arch is symmetric because it remains unchanged if left and right are reversed about a vertical line through the centre. The laws of electricity are symmetric under reversal of positive and negative charge. And so on.

The symmetries that underlie the revolution in our understanding of the four forces are of a very particular type, known as gauge symmetries. In Chapter 4 some simple examples were given of these abstract symmetries, such as the invariance of the laws of mechanics to changes in the zero level of height. Gauge symmetries are connected with the idea of re-gauging the level or scale or value of some physical quantity, and a system possesses a gauge symmetry if the physical nature of the system remains unchanged under such an alteration. Let us now see, with the aid of a simple example, how the abstract notion of making a gauge change can relate to the more concrete idea of a physical force.

Imagine being in a spacecraft out in the depths of space, far away from any planet or other astronomical body, travelling at constant speed in straight-line motion. You feel no forces at all, no sense of motion. You are completely weightless, floating freely. This is an easy image to visualize.

We now want to make a 'gauge change' to this scenario. That is, we want to alter the description by re-gauging or re-scaling a quantity, in this case a distance. Suppose the spacecraft is still travelling at constant velocity way out in space, but along a parallel track 1 km from the original. What would such a gauge change mean for the passenger? It would mean nothing, at least as far as forces are concerned. The passenger would feel exactly the same as in the previous scenario. More precisely, the behaviour of physical objects around him is completely independent of which straight line path of motion the craft is following. There is clearly a symmetry involved in this example. It could be expressed by the statement that the laws of physics remain invariant under a translation (i.e. shift) in the gauging of distance. So far, forces have not really entered the picture.

In making the gauge change, the path of the spacecraft remained a straight line. The shift in distance involved was *the same* at all points along its track. Expressed differently, the gauge change was the same everywhere, a concept known to physicists as a 'global' gauge transformation. The global character is important: if the re-gauging were to vary continually along the path of the spacecraft, the re-gauged path would then be a wiggly line. A spacecraft programmed to follow such a track would have to keep firing its motors, and as it zig-zagged violently around the passenger would be hurled about. He would feel forces. The behaviour of physical objects around him in the spacecraft would be distorted by the manoeuvering. Gauge changes that vary from place to place are known as 'local' gauge transformations. Quite clearly, the laws of physics are not invariant under local gauge transformations, which curve the path of the spacecraft and make the passenger feel uncomfortable. Or are they?

To take a simple case, suppose after the re-gauging that the spacecraft is programmed to fly in a circle at constant speed. The astronaut feels the curvature in the path of motion because he is no longer weightless. He does not float freely any more. Instead, he is pressed against the wall of the craft by centrifugal force. The physics within this circularly moving capsule is very different from the physics in a capsule moving uniformly in a straight line.

Imagine now that you are this astronaut, going round and round in empty space. You fall asleep, and when you wake up you are weightless once more. 'Ah', you think, 'the spacecraft must have returned to straight flight.' You look out of the window and to your puzzlement you see the stars going around. How can you be weightless yet still be moving around in a circle? A glance out of the window on the opposite side of the spacecraft reveals the reason why: you are in circular orbit around a planet.

One of the more entertaining spectacles of real spaceflight is when astronauts float about weightlessly when in orbit around the Earth. Their experiences in this condition are indistinguishable from those of an astronaut who is zooming away through interstellar space on a straight path to the stars at fixed speed. Here is a deep principle of nature: physics in a curved path around a planet is the same as in a straight path in deep space. And the reason is clear: the *gravity* of the planet exactly *neutralizes* the effects of curving the spacecraft's path. Physicists say that gravity is a 'compensating field'; it can compensate precisely for the deviation of a system from straight line motion. Of course, we have chosen a simple example, namely circular motion. To compensate for a spacecraft moving on a wiggly line you would need a much more complicated gravitational field, but the point is that whatever path the spacecraft is assigned it is possible to conceive of a gravitational field which will restore its passengers' comfort and weightlessness. Gravity could, in principle, always be made to transform away the violent buffeting of an erratic trajectory.

A conclusion follows from this story, and it is a very profound conclusion indeed. The laws of physics can be made symmetric even under *local* gauge transformations in distance provided a gravitational field is introduced to compensate for the place-to-place variations. Physicists like to turn this statement around and say that the gravitational field is nature's way of maintaining a local gauge symmetry, a freedom to re-gauge the scale of distance arbitrarily from place to place. In the absence of gravity, there is only a global symmetry; we can transform only from one straight line path to another without distorting the physics. With gravity, we can transform to any shape of path whatever without changing the physics. A symmetry, remember, is the invariance of something under an operation. The symmetry involved here is the invariance of physics under arbitrary changes in the shape of a path of motion. Viewed this way the force of gravity is simply a manifestation of an abstract symmetry – a local gauge symmetry – that underlies the physics of the world. It is as though the Great Designer said to himself, 'I do so like the beauty of symmetry. How nice it would be to have gauge symmetry at each place. So be it. Ah! I see I have invented a new field. Let it be called gravity.'

The power of the gauge symmetry concept is that not just gravity but all four forces of nature can be generated in this way. They can all be regarded as 'gauge fields'. In a *quantum* description of gauge fields coupled to particles of matter the concept of a gauge change must be widened further and related to the phase of the quantum wave which describes a particle. The technical details need not get in the way here. The essential point is that nature exhibits a number of local gauge symmetries and is compelled to introduce several force fields to compensate for the gauge changes involved. The force fields can be regarded as nature's way of imposing local gauge symmetries on the world. From this point of view, the electromagnetic field, for example, is not simply a particular sort of force field which just happens to exist. It is actually a manifestation of the *simplest known* gauge symmetry that is consistent with the principles of special relativity. The gauge transformations in this case correspond to changes in 'voltage' from place to place.

It is intriguing to think that a theoretical physicist who knew nothing of electromagnetism, but who wisely believed that nature is sold on symmetry, could deduce its existence entirely from the twin requirements of the simplest local gauge symmetry and the so-called Lorentz–Poincaré symmetry of special relativity that we met in Chapter 4. Using only mathematics, and guided by these two powerful symmetries, the theorist would be able to reconstruct Maxwell's equations without ever conducting an experiment involving electricity or magnetism, or even suspecting their existence! He would probably reason that, as the symmetries involved were of the simplest and most elegant form, it would be unusual if nature had not availed herself of them. By this purely abstract reasoning our theorist would be led to expect that electromagnetism existed in the real world. He could even go on to work out all the laws of electromagnetism, the existence of radio waves, the possibility of the dynamo motor, etc., that in reality were discovered using practical experiments. Such is the power of mathematical analysis in the description of the world that it can lead us to guess features that we may never have actually seem or previously suspected.

The concept of gauge symmetry turns out to be far more than a pleasing mathematical elegance. It is the key to constructing quantum theories of the forces that are free from the destructive infinite terms discussed in the

8

The Grand Trinity

A new force

Historians will look back on the 1970s as the time when scientists discovered that there weren't four forces of nature after all. The electromagnetic force and the weak force, superficially very different in character, are in fact two parts of a single force, called the electroweak force, which had never been suspected to exist.

The unification of these two force fields marked an historic advance on the road to a superforce. The first step was taken by Maxwell over a century before when he combined electricity and magnetism. The electroweak theory in its final form was largely the work of two men working independently, Steven Weinberg of Harvard University and Abdus Salam of Imperial College, London, though they built upon earlier work by Sheldon Glashow. Their theory strongly influenced the shape of particle physics in the years that followed.

The essence of Weinberg and Salam's theory is a description of the weak force in terms of a gauge field. This step must be taken first before there is any hope of unification. In the last chapter we saw how the concept of gauge symmetry is the key element in formulating a theory of forces free from the problems of infinite terms.

To treat the weak force as a gauge field theory we must believe that all particles which feel the weak force act as sources of a new type of field – a weak force field – although this field is not noticed by us directly. Weakly interacting particles, such as electrons and neutrinos, carry a 'weak charge' analogous to electric charge, which couples these particles to the weak field.

If the weak force field is to be regarded as a gauge field (i.e. as a way for nature to compensate for local gauge changes of some sort) the first step is to discover the precise form of the gauge symmetry involved. We have seen that electromagnetism already lays claim to the simplest gauge symmetry. It is no surprise that the weak force has a more complicated symmetry structure than the electromagnetic force because it acts in a more complicated way. In the example of neutron decay the weak force involves no less than four different types of particles (neutron, proton, electron, and neutrino), and the operation of the force leads to identity transformations among these particles. In contrast, the electromagnetic field leaves the identities of the source particles unchanged.

This tells us that the weak force embodies a more elaborate gauge symmetry connected with changes in particle identity. We encountered a gauge symmetry of this very sort at the end of Chapter 4. It went under the name of 'isotopic spin symmetry'. As originally developed, isotopic spin symmetry was supposed to describe the strong nuclear force between protons and neutrons. Remember that a gauge change in this case corresponded to turning an imaginary knob that mixed together the identities of protons and neutrons. The idea was that nuclear forces are invariant under such imaginary transformations. Weinberg and Salam, in their description of the weak force as a gauge field, borrowed this isotopic spin symmetry idea from the realm of nuclear physics and adapted it to the completely different topic of the *weak* interaction. The same essential notion of an identity-mixing symmetry was employed, but the hybridization used merges the identities of weak force sources, such as electrons and neutrinos.

Imagine a magic knob which enables us to change electrons into neutrinos and vice versa. As the knob is turned so the electron-ness of all electrons gradually fades until they turn into neutrinos. At the same time, the neutrino-ness of all neutrinos fades and they turn into electrons. This can't actually happen, of course, but theorists can explore the consequences of such fictitious procedures by examining the equations which describe the particles and forces.

What has just been described is an example of a global gauge change. It is global because turning the knob alters the identity mix of every electron and every neutrino in the universe. In the previous chapter we saw how going from a global gauge symmetry to a local one leads to the invention of force fields, which are needed to compensate for the gauge changes that differ at each point. Re-gauging the definition of position at each point served to introduce the concept of a gravitational field. The global gauge change described by the magic knob could also be converted into a local one. We could imagine a separate knob for each point in space, and our setting the positions of all the knobs differently. When this is done, new force fields are needed to maintain symmetry, to compensate for the chaotic re-settings of the knobs from place to place. These new force fields turn out to describe precisely the weak force. The fact that the gauge symmetry involved here is rather more complicated than the electromagnetic case is reflected in the fact that three new force fields are necessary to maintain symmetry. This is in contrast to a single electromagnetic field. A quantum description of these three fields can readily be given: there will be three new types of messenger particle, one for each field.

How well do these three fields describe the weak force? The purpose of the fields is to compensate for place-to-place variations in the identity mix of electrons and neutrinos. (The theory can also be used for other leptons or for quarks.) This means that when a quantum of the field is emitted or absorbed, there will be an abrupt change in the identity of the particle. An electron may change to a neutrino or vice versa. This is exactly what happens when the weak force acts.

A typical weak force process is shown in Figure 17. The experimenter

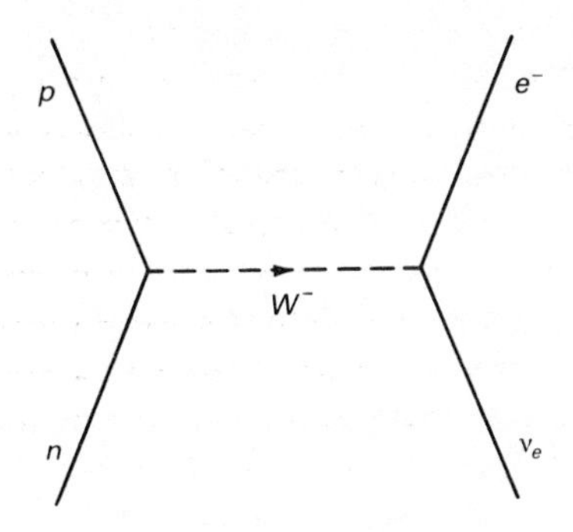

Figure 17. In this weak force process a neutron (n) encounters a neutrino (ν_e), and they transmute into a proton (p) and an electron (e^-). Careful examination reveals that this effect is produced by the exchange of a heavy, charged messenger particle (W^-) from the neutron to the neutrino. The W^- brings about the transmutation of a neutron into a proton by changing the flavour of one of its quarks from down to up at the instant of its emission.

sees this event as a scattering event in which a neutron (n) encounters a neutrino (ν_e), and the two particles undergo identity changes to produce a proton (p) and an electron (e). A more detailed description, in terms of messenger particle exchange, has a down quark in the neutron turning into an up quark (thus changing the neutron into a proton) with the emission of a messenger (represented by the broken line in the figure) that is subsequently absorbed by the neutrino, transmuting it into an electron. Because the proton appears with positive electric charge, the messenger particle must carry with it negative charge (by the law of charge conservation). This negative charge ends up on the electron. The negatively charged messenger particle is called a W^-. There will also be a positively charged antiparticle, W^+. The W^+ could transmit the weak force from, say, an antineutron to an antineutrino.

The W^+ and W^- account for two out of the three weak force fields predicted by the Weinberg–Salam theory. The remaining field corresponds to an electrically neutral messenger particle, dubbed the Z. When this theory was first formulated the idea of a neutral messenger for the weak force was novel. If Z existed, it would show up as a weak force between particles that would not involve any transfer of electric charge. An example is shown in Figure 18. Here an electron and a neutrino scatter from each

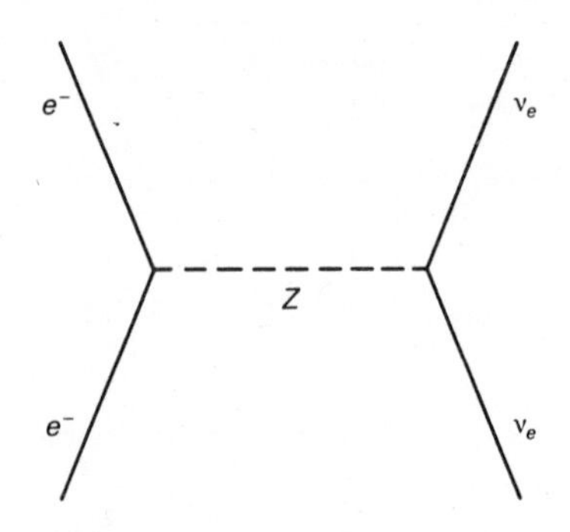

Figure 18. The theory of Salam and Weinberg predicts that electrons can scatter neutrinos by the exchange of a new type of electrically neutral weak messenger particle, the Z. Experimental evidence for such processes was obtained in the mid-1970s.

other by the exchange of a Z. In 1973, a long-running experiment at CERN demonstrated that electrically neutral weak forces really do exist. The result was a powerful boost for the Weinberg–Salam theory.

In spite of this happy accord between theory and observation, the description of the weak force as a gauge field still had to surmount a major obstacle. It is in the very nature of gauge fields that they are long-ranged, and the theory seems inevitably to lead to the prediction that the messenger particles are massless, like the photon. In reality the weak force is very short-ranged, and its messenger particles carry a huge mass. If the Ws and Zs are simply assigned a mass in the theory, the crucial gauge symmetry is destroyed. How is it possible to get the best of both worlds, i.e. gauge symmetry as well as messengers with a mass?

The resolution of this conundrum was provided by Weinberg and Salam in 1967. It involved a clever idea known as 'spontaneous symmetry breaking'. This is how it works.

Imagine a smooth surface shaped like a Mexican hat resting horizontally (*see* Figure 19). A small ball is placed at the centre of the hump. With this arrangement the system has an obvious symmetry, namely that it is unchanged by rotating around a vertical axis passing through the centre of the hat. As far as the force of gravity is concerned, there is no preferred horizontal direction (gravity acts vertically); one place of the rim is equivalent to every other.

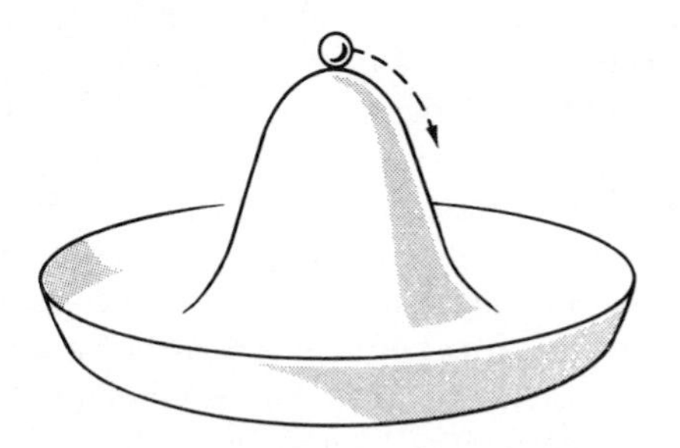

Figure 19. Spontaneous symmetry breaking. The ball is placed on top of the 'Mexican hat' surface. In this configuration there is complete rotational symmetry. However, the configuration is unstable, and the ball spontaneously rolls down into the rim of the 'hat', coming to rest at some arbitrary point. The rotational symmetry is thereby broken. The system has traded symmetry for stability.

Although the system is symmetric, it is not stable. Once the ball is released it will not remain poised for ever on top of the hump, but will soon roll down the slope and come to rest somewhere in the rim of the 'hat'. When this has happened the symmetry will have been destroyed. The ball will have picked out a particular place on the rim to stop, thereby defining a privileged horizontal direction from the central axis. Stability will have been traded for symmetry. In the stable configuration the underlying rotational symmetry of the forces (in this case gravity) will still exist, but it will be hidden. The actual state of the system will no longer reflect the symmetry of the forces which control it.

This same general idea was used by Weinberg and Salam, though the symmetry involved was a gauge symmetry rather than a rotational symmetry, and the spontaneous breaking corresponded to the quantum state adopted by the force fields. Thus, in their theory the underlying gauge symmetry is still present in the fields, but the fields cannot normally exist in a state which reflects this symmetry, because such a state is unstable. Therefore the field 'sinks' into a stable state which breaks the symmetry and bestows upon the messenger particles a mass. Naturally, the details are more complicated than the 'Mexican hat' example, but the basic idea is the same: the symmetry is still there in the underlying laws but is not reflected in the actual state of the system. For that reason physicists did not spot this vital gauge symmetry during thirty-five years of study of the weak force.

To achieve the crucial spontaneous symmetry breaking Weinberg and Salam introduced into the theory an additional field, called a Higgs field after Peter Higgs at the University of Edinburgh, who had already studied spontaneous symmetry breaking in the context of particle physics. Nobody has ever seen a Higgs field, but its presence can have a crucial effect on the behaviour of the gauge fields. In the case of the 'hat', the symmetric state with the ball placed at the top is unstable. The ball prefers to roll into the rim because the broken symmetry state has a lower energy. Likewise, the behaviour of the Higgs field is such that its lowest energy state is one of broken symmetry. It is the coupling between the Higgs field and the gauge fields that gives the W and Z particles a mass. The theory also predicts the existence of a Higgs particle – a quantum of the Higgs field – that has zero spin and a large mass.

Having seized upon the idea of spontaneous symmetry breaking, Weinberg and Salam were then able to take the next momentous step, and to combine electromagnetism and the weak force into a single gauge field theory. To enable both sorts of field to emerge from a unitary theory, it was necessary to begin with a more elaborate gauge symmetry, one which embodied both the simple gauge symmetry associated with electromagnetism and the 'isotopic spin' symmetry associated with the weak force. There are thus four fields present in all in this theory, the electromagnetic field and three weak force fields. The next step was to arrange for Higgs fields to bring about a spontaneous symmetry breaking. Initially the W and Z quanta are massless, but the effect of the symmetry breaking is for some of the Higgs particles to coalesce with the W and Z particles and endow them with a mass. As Salam puts it, the W and Z particles eat the Higgs particles in order to gain weight. The photon is left untouched by this process and remains massless.

The Weinberg–Salam theory explains beautifully why the electromagnetic and weak forces appear to have such dissimilar properties. The underlying structure of their force fields is actually much the same; both are gauge fields. It is the effect of symmetry breaking that produces such a difference in their character. We fail to notice the gauge symmetry of the weak force because it is hidden from us by the symmetry breaking.

Another big difference between the two forces is their strength. Why is the weak force so weak? This too is explained by the theory. If the symmetry remained unbroken both forces would have comparable strengths. The symmetry breaking has the effect of enfeebling the weak force part. In fact, the strength of the weak force is directly related to the masses of the W and Z. It might be said that the weak force is so weak because the W and Z are so massive.

When Weinberg and Salam published their theory in the late 1960s one major theoretical challenge still remained. Was the theory renormalizable? Would the QED infinity-squashing miracle work for the combined electroweak gauge fields? This problem was taken up by Gerhardt 't Hooft of the University of Utrecht in the early 1970s. The task was a formidable one, involving complex and tedious computations of successive terms in a long sequence to see where the deadly infinities might lie. Some of the labour was relieved by use of a computer. 't Hooft later described how at the completion of the computations he anxiously examined the computer output:

> 'A few simple models gave encouraging results: in these selected instances all infinities cancelled no matter how many gauge particles were exchanged and no matter how many loops were included in the Feynman diagrams. The decisive test would come when the theory was checked by the computer program for infinities in all possible diagrams with two loops. The results of that test were available by July, 1971; the output of the program was an uninterrupted string of zeros. Every infinity cancelled exactly.'

Evidently the high degree of symmetry built into the electroweak theory was the crucial element in avoiding the catastrophe of the infinities. It was a lesson well learned.

All that remained was for a definitive experimental test of the new theory to be carried out. The most satisfactory confirmation had to be the positive identification of the still-hypothetical W and Z particles.

In a laboratory experiment in most circumstances the W and Z particles go unseen. They remain as virtual quanta, exchanged as messengers between other particles. If, however, enough energy can be pumped into the system, the Heisenberg energy loan which finances the fleeting existence of the W and Z particles can be cleared and these particles can then become 'real', i.e. they can fly off and lead an independent existence. Being so massive (about ninety proton masses), liberation of Ws and Zs in this way requires huge amounts of energy, and so it was only with the recent development of very large accelerators that it was possible for them to be produced and identified.

With the eventual discovery of W and Z particles in 1983, the Weinberg–Salam theory was triumphantly confirmed. No longer was it necessary to talk of four fundamental forces. The superficially separate electromagnetic and weak forces were seen to be merely two components of a single electroweak force. For their great accomplishment Weinberg and Salam received the Nobel prize in 1979, shared with Sheldon Glashow of Harvard University for laying the foundations of theory in earlier work.

Inspired by the brilliant achievements of the electroweak theory, physicists began to wonder whether further unification was possible. Perhaps there are actually only two fundamental forces of nature, or maybe even a single superforce? It was not long before the strong force came under closer scrutiny.

Coloured quarks and QCD

The idea that the strong force might be amalgamated with the electroweak force in a further unification was not long in coming once the success of the Weinberg–Salam theory was clear. But before any such unification could be implemented it was first necessary for the strong force to be cast in the form of a gauge field. We have seen how the strong force can be pictured in terms of the exchange of gluons which serve to bind together quarks into pairs or trios to form the so-called hadrons. A gauge field description of this process can be constructed by using once more the isotopic spin symmetry concept, suitably generalized.

The essential idea is as follows. Each quark possesses an analogue of electric charge which acts as the source of the gluon field. For lack of a better word this 'charge' is called *colour*. (No connection is intended, of course, with ordinary colour.) The electromagnetic field is generated by only one sort of charge, but the more complicated gluon field requires three separate colour charges. Each quark, then, can come in one of three possible colours, which are arbitrarily referred to as red, green, and blue.

The gauge symmetry associated with these colours can be visualized once again in terms of a 'magic knob' which mixes identities. In this case the knob has *three* pointers – red, green, and blue (*see* Figure 20) – rather than two. Rotating the knob corresponds to converting red quarks into green or blue, and so on, depending on the setting. Again, the conversion is continuous, with redness slowly fading into blueness, etc.

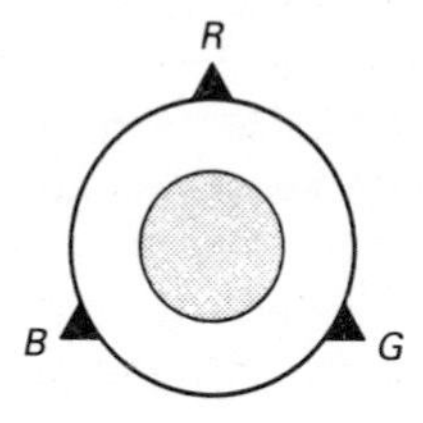

Figure 20. Magic knob equipped with three pointers represents the more elaborate gauge symmetry associated with quark colour. The interquark (gluon) force is unchanged by rotations of the knob, which merge the quark colours, red (R), blue (B) and green (G).

The theory now unfolds along the same lines as for the weak force. The demand for local gauge symmetry – invariance under independent colour variations at each point in space – requires the introduction of force fields to compensate. This time, because the knob has three pointers rather than two, the symmetry involved is more intricate, and this is reflected in the higher number of force fields that are necessary to maintain local gauge symmetry. In fact, a total of *eight* new compensating force fields must be invoked. The messenger particles of these fields are, of course, the gluons, and the analysis therefore implies that there should be eight different types of gluon. Contrast this with the single messenger particle of electromagnetism (the photon) and the three for the weak force (W^+, W^-, and Z).

Antiquarks come in three anticolours (antired, antigreen, and antiblue). The gluons themselves also carry colour, of a composite hue, such as blue–antigreen. Whenever a quark dispatches a gluon messenger its colour has to change to pay for the colour carried away by the gluon. Thus, a red quark can emit a red–antiblue gluon and change to blue. Similarly, a green quark which absorbs a blue–antigreen gluon changes to blue. And so on.

The effect of gluon emission or absorption, then, is to change the identity of a quark, for instance from a red quark to a green quark. In this respect the strong force resembles the weak force, for which the emission of a W changes, for example, an electron to a neutrino. Quarks are subject to the weak force too, as well as the strong force, but the change in identity that occurs when a weak messenger is emitted is different from that occasioned by the emission of a gluon. Whereas gluons change the *colour* identity of quarks, the weak force changes the *flavour*. For example, when a neutron decays, one of its down quarks emits a W^- and changes into an up quark. It is important to remember that quarks possess both colour and flavour, and the two should not be confused.

In a typical hadron such as a proton the three quarks are continually exchanging messenger gluons and changing colour. The changes are not arbitrary, though. The mathematics of this theory imposes a very important rule which must be followed unswervingly during this multichrome frolic. At any given time the 'sum' of all three quark colours must always be red + green + blue. Pushing the analogy with real colour a stage further, we can say that the combination of colours in a hadron must always make 'white' (merging the primary colours red, green, and blue does give white). This is the all-important gauge symmetry at work. The operation of the gluon fields compensates for internal changes in the quark colours so as to maintain the pure whiteness of the hadron.

Hadrons can also be made out of quark–antiquark pairs, the mesons. Because an antiquark carries anticolour, such a combination is guaranteed to be colourless ('white'). For example, a red quark combined with an antired quark makes a colourless meson. In this scheme all leptons are colourless too, because they do not feel the gluon force at all.

The quantum theory of colour – or quantum chromodynamics (QCD) as it is called – beautifully explains the rules of quark combination that were originally worked out on an *ad hoc* basis in the 1960s. From the standpoint of QCD the strong force is nothing more than nature's insistence on maintaining an abstract symmetry, in this case that all hadrons remain white even when internal colour changes occur. If we demand this abstract gauge symmetry of nature, the gluon fields are forced upon us. We do not need to invent them – they come automatically out of the mathematics.

There is one important feature of the strong force which has not yet been mentioned. When the quark theory was first invented, it seemed that a clear test of the theory was readily available. All that was needed was for someone to smash a hadron apart and display the constituent quarks for the world to see. An isolated quark is conspicuous because its electric charge is either ⅓ or ⅔ of the charge carried by every other sort of particle.

As larger and larger accelerators came into use and still nobody succeeded in breaking a hadron apart, physicists began to wonder if the quark theory might be wrong. Surely, if there are quarks inside a proton it must be possible to knock one of them out if the proton is given a hard enough blow? Yet even when struck with an energy equivalent to many times its own rest mass, a proton resolutely refuses to be disintegrated. All that happens in these events is for a whole shower of new, fully assembled hadrons to appear. Individual quarks are never seen.

An alternative strategy in the search for the quark is to fall back on nature. If quarks exist, it is reasonable to suppose that they will have been produced naturally somewhere. Presumably when matter was originally created, quarks were made first, and then they combined together to form hadrons. It might reasonably be conjectured that a few quarks would have failed to find partners and be left to roam the universe in isolation. A search of ordinary matter might then reveal the odd quark still lying around.

William Fairbank of Stanford University embarked on just such a search. He carefully examined small samples of natural minerals, such as niobium, to see if they contained any particles with electric charge of ⅓ or ⅔. This he did by observing the behaviour of the samples when placed in a strong electric field. Fairbank's very careful experiments were repeated over several years, and on a number of occasions he reported positive results. Fractionally charged particles were, he claimed, present on some of the samples. To this day Fairbank's claim remains open, but similar experiments conducted elsewhere have failed to confirm his results, and many physicists remain sceptical. Does this mean that, in spite of everything, the existence of quarks is in doubt? Not at all. There is a growing feeling that quarks can exist *only* inside hadrons. If so, there must be some law of nature which forbids isolated quarks to exist. Every time you try to pull a quark out of a hadron, something must stop you removing it completely. That something is presumably the gluon force field. Evidently, the quarks are bound so tightly inside hadrons that no force in the universe can break the bonds and free them. Physicists say that the quarks are 'permanently confined' inside hadrons and refer to the sought-after explanation for that fact as the confinement problem.

Superglue

A major theoretical challenge is to understand quark confinement in terms of gauge field theory. If an individual quark could be produced, it would display a particular colour charge – red, green, or blue. While quarks are confined, we never see colour, only the overall 'white', or colourless, combinations. If confinement is permanent, then for some reason nature does not permit us to observe naked colour. It is censored. That would explain why leptons but not quarks can exist in isolation, since leptons are colourless.

But what happens if we simply try to pull a quark out of a hadron? What sort of superglue is it that holds on so tenaciously it can never be overcome?

One important clue about the nature of the force between quarks came from the SLAC experiments, mentioned in an earlier chapter, in which very high-energy electrons were shot through protons. The results indicated that at short distances the force more or less fades away and the quarks behave essentially as free particles. More information can be gleaned from the behaviour of mesons where a quark and an antiquark form a bound system reminiscent of a hydrogen atom. By studying the excited states of a hydrogen atom it is possible to deduce the inverse square law of electric force between the proton and electron in which their mutual attraction declines steadily with separation. Similar studies of the excited states of mesons point to a completely contrary situation. As the two particles move farther apart to higher energy states, the force between them becomes stronger rather than weaker.

These results suggest that the force between quarks is very odd. All the other forces of nature become weaker with distance. The inter-quark force does the opposite. It has been likened to a piece of elastic which pulls harder the more it is stretched, but goes loose when the ends are close together. Another analogy is with a chain. It is as though the quarks are chained together inside hadrons. If they keep close the chains are not really noticed and the quarks are free and independent within a narrow range. But if one of the quarks tries to break away it gets pulled up sharply. Physicists refer to this arrangement as quark 'slavery'.

Of course, once the idea of quark slavery became fashionable the thought in everybody's mind was whether QCD could provide an explanation for it. In the event the calculations turned out to be immensely difficult, although they have provided a number of encouraging pointers. From a physical standpoint it is possible to understand in outline how it might come about that the inter-quark force grows larger with distance.

A basic difference between electromagnetism, where the force diminishes with distance, and the gluon field is that photons do not carry electric charge. If they did the world would be altered beyond recognition. By contrast, gluons carry colour 'charge' in hybrid combinations, such as red–antigreen. But colour is the source of the strong force. Therefore not only do gluons serve to stick quarks together, they also tend to stick to each other. This makes things rather complicated, but a careful analysis suggests that the universal stickiness of the gluons could be the key to explaining quark slavery.

To see how, we have to go back to the concept of a quantum vacuum. Let us look first at what happens to an electron when it is placed in the vacuum. Remember that the space around the electron is not really empty, but filled with virtual particles of all varieties. Among these will be virtual electrons and virtual positrons. Although we don't see these virtual particles directly, we know that they are there and that they can leave physical traces. The electron we have placed in the vacuum will also know they are there because they will react to its presence. The electric field of the electron will disturb the pattern of activity of these virtual electrons and positrons during their brief existence. The positrons will tend to be drawn towards the electron of interest by electric attraction, while the virtual electrons will be repelled. There is thus a net shift of charge, a phenomenon known as polarization. The fact that empty space can become electrically polarized in the presence of an electric field is a curious consequence of quantum theory. It is hard to imagine a vacuum having electrical properties, but the effect is nevertheless real and has actually been measured in experiments.

The upshot of this vacuum polarization is that a sort of neutralizing screen of electric charge forms around the electron. There is no way the electron can rid itself of this screen; it is part of the shroud of virtual particles that all electrons carry round with them. From a distance, the effective charge that an electron appears to possess is less than the charge it really has because of the screening effect. If we could probe inside the shroud we should begin to perceive the 'bare' electron, with its much greater charge. We should find as we penetrated the cloud that the simple inverse square law of force that holds at a distance begins to fail because of the cloud of virtual positrons that hugs the central electron. Thus, the presence of vacuum screening can alter the way that a force varies with distance.

Screening also occurs in the gluon field, where the effect is to modify the strength of the colour 'charge' carried by a quark. Virtual antiquarks will tend to cluster round a quark of the appropriate colour. For example, a red quark will attract a shroud of antired antiquarks. As in the electromagnetic case, the result is a partial neutralization of the colour charge. This time, however, there is an additional contribution to the vacuum polarization arising from the gluons. Because gluons also possess colour, virtual gluons in the vacuum will respond to the presence of a quark. It turns out that the gluon shroud acts in the opposite manner to the quark shroud, tending to reinforce rather than neutralize the colour on the central quark. The effect of the virtual gluons is therefore to oppose the action of the virtual quarks, and detailed calculations confirm that the gluons win out, the net result being that the colour charge of the quark is *enhanced* by the vacuum rather than diminished by it.

The consequences for the inter-quark force are profound. If a neighbouring quark penetrates the virtual particle shroud, it finds that the effective colour of the quark within the shroud diminishes, and the force dwindles away. The intruding quark is released from the grip of the other quark so long as it remains ensconsed within the shroud. Close up, then, the inter-quark force is emasculated. This is directly opposite to the situation in electromagnetism, and precisely the right general behaviour to explain quark slavery.

It would be premature to pronounce QCD today the unqualified success that QED was hailed on its completion some forty years ago. Nevertheless, it is an impressive advance. Even in the 1960s the physics of hadrons seemed to be a tangle of complicated forces and incomprehensible particles. QCD has unravelled this mess and provides a simple base on which a theory of hadrons can be built with relatively few parameters.

Grand unification

With the inception of QCD, all the forces of nature at last possessed a common description in terms of gauge fields. This brought with it a new hope. The successful unification of the weak and electromagnetic forces within the framework of gauge field theory suggested that further unification was possible. In 1973 Sheldon Glashow and Howard Georgi published a theory in which the new electroweak force was merged with the strong gluon force to form a 'grand unified force'. This was the first

GUT, or grand unified theory. There are now several contending GUTs but they are all based on the same essential idea.

If the electroweak and strong forces are really just two facets of a single grand force, then the grand force must be described as a gauge field with a symmetry that is sufficiently elaborate, or 'large' enough, to embody the gauge symmetries already contained in QCD, and the Weinberg–Salam theory. This is a mathematical job. There is no unique symmetry which will do the trick, hence the proliferation of rival theories. Nevertheless, all GUTs have a number of features in common.

One of these is that quarks, the carriers of the strong force, and leptons, the carriers of the electroweak force, become amalgamated into the same theoretical structure. Quarks and leptons had hitherto been regarded as entirely separate beasts, and their incorporation into a single theory was a totally novel concept. It marked a further important step on the road to unification.

The gauge symmetries involved in GUTs can also be thought of in terms of a 'magic knob' that mixes the identities of particles, but this time the number of pointers has grown again. Instead of the two needed for the electroweak force and the three needed for QCD, there are five pointers. Turning the GUT knob, therefore, is a powerful business. It can do things the other lesser knobs were forbidden to do, such as change quarks into leptons, or even into antiquarks, processes that were ruled absolutely out of court in previous theories.

As before, the demand for nature to respect an abstract gauge symmetry – a bigger one this time – by conjuring up compensating force fields leads us to discover, mathematically, new types of fields with new properties, such as the ability to turn quarks into leptons. In the simplest GUT due to Georgi and Glashow, the magic knob connects together red, green, and blue down quarks, a positron, and an antineutrino. This requires twenty-four unified force fields. Twelve of the quanta of these force fields are already known: the photon, the two Ws, the Z, and eight gluons.

The remaining twelve are new, and are given the collective name X. These correspond to the fields whose job it is to maintain the larger gauge symmetry which mixes the quarks with the leptons (corresponding to knob settings which hybridize, say, a red down quark with a positron). Therefore the quanta of these fields – the X particles – can change quarks into leptons, or vice versa, when they are exchanged as messengers. X particles carry electric charges of ⅓ or ⁴⁄₃.

Let us now trace the fate of a typical hadron which experiences the activities of these unusual X particles. A good subject is the proton, which contains two up quarks and one down quark. The down quark mixes with the positron via the grand gauge symmetry, and can transmute into it by exchanging an X of the appropriate qualities, in this case one which has the relevant colour charge and an electric charge of $-4/3$. The X must, of course, be conveyed to some other particle, which could be one of the up quarks in the proton. The recipient up quark will absorb the X and thereby turn into an anti-up antiquark.

Something rather remarkable has happened here. What started out as a triplet of two up quarks and one down quark has transmuted into a positron and a quark–antiquark pair. The positron does not feel the strong force, and so it flies off out of the hadron on its own, while the remnant quark–antiquark pair no longer constitutes a proton, but a meson – in fact, a pion. To an experimenter, this sequence of events would therefore appear as the *decay of a proton* into a positron and a pion.

In the entire history of particle physics there has always been an inviolable rule that the proton is an absolutely stable particle. After all, ordinary matter is made of protons. That the proton might be unstable and likely to decay is a dumbfounding prediction of GUTs. It implies that all matter is ultimately unstable, and therefore impermanent, a very profound conclusion indeed.

In the early winter of 1974 I drove from London to a conference at the Rutherford Laboratory near Oxford. One of the passengers I took along was Abdus Salam, and we started talking about the content of the lecture he would give. Salam said he had some ideas about the possibility that protons might decay. I remember being bemused by the whole business, and not a little sceptical. Salam duly gave his lecture, and most prophetic it turned out to be, but the conference was remembered instead for the talk by Stephen Hawking of Cambridge University, who announced his sensational discovery that black holes are unstable, and eventually explode amid a shower of radiation. Curiously, the Hawking process can also cause protons to decay, as he realized some years later. Quantum effects permit a proton to spontaneously 'tunnel' inwards to become a virtual black hole, which then evaporates by the Hawking process, disgorging a positron on the way. The black hole demise of a proton is much less probable, however, than the GUTs route.

Obviously the abrupt disappearance of a proton is an event likely to attract the attention of a sharp-eyed particle physicist, and so the question arises as to why proton decay was not discovered long ago. To deal with this question it is necessary to inquire into the *rate* of decay predicted by the theory. Experience of radioactivity shows that half-lives for transmutation can vary enormously, depending on the strength of the interactions which drive the decay and the masses of the particles involved. A crucial parameter in the case of proton decay is the mass of the X particle, which according to the rules of the game determines its range. If X is very massive, its sphere of activity will be severely limited. For proton decay to occur, the two participating quarks have to come close enough to swap an X, and chance close encounters are very rare. This would explain why proton decay was not already known to experimenters. Using the best available estimates for the half-life of the proton and working backwards, the mass of the X comes out at about 10^{14} proton masses, a colossal value against which the heaviest known particle to date – the Z, with a mass equivalent to about 90 protons – pales into insignificance.

Before dwelling on this dizzying number, an apparent paradox must be resolved. The reader may be baffled as to how a proton can contain within itself messenger particles that are 10^{14} times heavier than it is. The answer is provided by the Heisenberg uncertainty principle. The X exists, remember, for the most minute duration, only while it is being exchanged between quarks which happen to brush very close to each other. For such a brief interval energy, hence mass, will have an enormous uncertainty.

The quantum theory links together energy (or mass) and distance through the uncertainty principle. Hence, a scale of mass automatically defines a scale of distance. The physics which is important at a certain distance is the physics which is important at a certain energy (or mass). That is why you need very high-energy accelerators to probe very small distances. The mass scale of X thus gives us a related distance scale, which is roughly the distance that the X travels in its role as a messenger. The actual distance works out at about 10^{-29} cm. This is the range of the X, and it tells us how close two quarks must come to exchange an X so that a proton will decay. Size for size, 10^{-29} cm is to a proton what a speck of dust is to a solar system. The world of GUTs and proton decay is millions of millions of times smaller than the world of quarks and gluons probed by our accelerating machines so far. It is like a universe within the proton, as inaccessible for us in its smallness as are the extragalactic spaces in their remoteness. To probe such a Lilliputian domain directly we should have to build an accelerator larger than the solar system.

Any complete theory of the forces of nature must explain the relative strengths of the different forces. GUT workers were quick to point out how their theories could account for the great disparity between the electroweak force and the strong force. The real strength of these forces is not what the experimenters measure when they observe the antics of subatomic particles because the sources of the various fields are all screened by vacuum polarization effects as already explained. The way this screening works is that the electromagnetic force gets stronger at short range, while the strong force gets weaker, and so there is a tendency to converge. The weak force, when its strength has been adjusted to allow for symmetry breaking, comes out in between the other two, and like the strong force it is 'antiscreened', and so it too gets weaker closer in. An interesting calculation is to determine at what distance the strengths of the three forces come together at the same value. The answer is about 10^{-29} cm once again, precisely the length scale associated with the X mass. It is a pleasing consistency.

The upshot of this analysis is that, at ultra-high energy (or, equivalently, at minute distances), the electromagnetic, weak and strong forces merge into a single force, and the separate identities of the quarks and leptons is lost. We only perceive distinct forces and particles in the world of our experience because we are examining matter at comparatively low energy. Physicists call 10^{14} proton masses 'the unification scale'. There is a dramatic meaning behind this number.

Before pursing this further, let us assess the status of the grand unified theories. By amalgamating three forces into one grand trinity, the GUTs greatly reduce the number of arbitrary parameters in our description of nature. The less ambitious Weinberg–Salam theory contains constants which must simply be fixed by experiment. In the GUTs some of these numbers are no longer chosen *ad hoc*, but are determined by the theory.

There are also some unexpected bonuses. One of these is a possible explanation for the age-old mystery of why electric charge always comes in multiples of the same fixed fundamental unit. Taking the convention that an electron has charge -1 and a proton $+1$, the down quark has the smallest unit of charge, $-1/3$. All other charges are small multiples of this value, whether found on quarks, leptons, or messengers. Before GUTs there was no known reason why particles could not exist with any value of charge, even values such as π. In the unified theory, however, this is not allowed. Rigid rules are imposed by the fact that all the particles belong to large family groups which can swap messenger particles carrying fixed units of charge. For example, in proton decay when the down quark turns into a positron and sheds a charge $-4/3$ on the X, the up quark which absorbs it must have the right charge to make up the correct antiquark charge after it absorbs the X. The arithmetic has to be consistent, which means all the particles in the family have to possess charges that are simple multiples of each other (or no charge at all).

Ranged against GUTs, however, is the fact that there is no unique theory, and the unification scale is so remote there is no prospect whatever that it will become accessible to direct experimentation. How, then, are we to discriminate between rival theories? If the GUTs describe a world so small and so energetic that we can never observe it, has not physics degenerated to pure philosophy? Are we not in the same position as Democritus and the other Greek philosophers who mused endlessly about the shapes and properties of atoms without any hope of ever observing them?

Some physicists fervently hope not, and they point to three lifelines that could yet give us a hold on physics at the unification scale. We shall take a look at each in turn.

9
Superforce Glimpsed?

Proton decay

The week that CERN announced the discovery of the Z, Steven Weinberg was attending a Royal Society meeting in London. He said he was rather depressed about the future of fundamental physics. This might well seem astonishing. How could Weinberg, whose own theory was that very week being spectacularly confirmed, be depressed about the way physics was going?

Like all brilliant theoretical physicists, Weinberg was already several jumps ahead of the experimenters. His interest had long ago gone beyond electroweak theory to grand unification, and further. Weinberg's depression was caused not by the splendid positive result from CERN, but by a less publicized negative result from Lake Erie.

The Lake Erie experiment was one of several repeated across the world that offer the best, perhaps the only, hope of testing GUTs, and glimpsing a pale shadow of physics at the unification scale. All these experiments have one aim – to spot a single proton in the act of disintegration. The sensational prediction of GUTs that protons are unstable came as something of a bombshell for most physicists, although as I have explained the idea had been around in the wings for a number of years. Experiments had long been conducted to establish limits on the lifetime of protons, but they were all of a routine nature. Nobody seriously expected to find a proton decaying.

When it emerged from GUTs that the unification mass scale was about 10^{14} proton masses, it was instantly clear that physicists would never get to explore grand unified physics by direct experimentation. The only hope of a test would be some indirect effect of GUTs, and proton decay was the obvious choice. The precise lifetime of the average proton predicted by the theory depends on which version of GUTs you take, but most of them, including the simplest (known cryptically as 'minimal SU(5)') gave a figure of around 10^{31} years. This is 10^{21} times the present age of the universe.

How can a process that takes incomparably longer than the age of the universe ever be observed by man? The answer lies in the fact that not every proton takes 10^{31} years to decay. The rules of quantum physics demand that each individual decay event is unpredictable; it is the average lifetime which works out at 10^{31} years, which means that if you assemble 10^{31} protons you can bet fairly confidently that one of them will decay in a year or two. This is the strategy adopted in all the proton decay experiments. Many tonnes of matter are taken somewhere well away from cosmic rays and monitored continuously for a sudden, single decay event. The cosmic rays are a major nuisance because they swamp the delicate instruments with all manner of unwanted particle events. To escape this pollution the experiment has to go under a mountain or down a deep mine. Even then you can't keep the neutrinos out.

Prior to GUTs, the best limits on the proton's lifetime were 10^{28} years. This is an impressive number, even though it refers to a non-event. It must be the best-attested example of something known to man not happening. What else is definitely known from direct experiment *not* to occur for at least 10^{28} years? With the inception of GUTs, proton lifetime measurements received a major boost. To verify the GUTs prediction the experimental accuracy had to be approved by a factor of at least a thousand, and this meant a more elaborate and expensive experimental set-up.

The prospect of actually detecting proton decay was so exciting to physicists that several research groups joined the race. Among the first off was a joint Japanese and Indian team who constructed a sort of layer cake of iron slabs deep in an Indian mine. Enveloping this superficially inert lump of matter was a cordon of particle detectors, tuned to spot the decay products of a proton. In the early spring of 1982 an announcement was made. Several 'candidate events' had been observed that suggested protons were decaying with a lifetime of about 10^{31} years, as predicted by the simplest GUT.

The claim was greeted with great interest but with a certain caution. Even at the depths at which the experiment was located cosmic rays and neutrinos could still mimic the effects of proton decay. Other experiments would be needed to confirm the results before physicists could be sure. A few months later a CERN experiment under Mont Blanc also recorded a likely-looking event, and interest began to grow. It seemed that we were tantalizingly close to the threshold of a new domain in physics.

A great deal of attention was given to some of the consequences that would follow if the proton really is unstable. Protons are the building blocks of all nuclear matter. If they are all destined to decay it means that, eventally, the universe will vanish. It won't happen suddenly, of course. Gradually, over the aeons, all matter will inexorably evaporate away. If the 10^{31} year estimate is correct it means that during your lifetime there is a good chance that at least one proton in your body will disappear.

What about the other constituents of atoms, the neutrons and electrons? The same process that causes proton decay can also destroy neutrons, though many would succumb to the more conventional beta decay process. Every proton that decays leaves its electric charge on a positron, which is the antimatter counterpart of the electron. Each emerging positron will therefore seek out and destroy an electron. As there are the same number of electrons as protons in the universe at the outset, it is probable that electron–positron annihilation will finish off virtually all the electrons in the universe. The end result, then, is that the 10^{50} tonnes of matter in the visible universe will one day be reduced to nothing. It is a sobering thought.

To be sure of this scenario, physicists had first to check the early claims that proton decay was a reality. More refined experiments were set up all around the world. One of the best made use of a salt mine, 600 m under Lake Erie. The protons for the experiment came in the form of 8 000 tonnes of highly purified water contained in a cubical tank some 18 m across. Dangling in the water were 2 000 photomultiplier tubes. These had the task of detecting the minute pulses of light which are produced when fast-moving, charged particles travel through a dense medium. The intention was to spot the energetic decay products from decaying protons by recording these brief flashes of light. If the lifetime estimate of 10^{31} years was correct, the Lake Erie experiment should have registered several events in the first three months of running. As it happened, not a single proton decay was seen. It began to look as if the earlier reports had been mistaken, and hopes started to fade that protons would ever be seen to decay.

This negative result does not disprove GUTs, but it does seem to rule out the simplest versions of grand unification. More complicated theories exist which predict much longer proton lifetimes, but it is then unlikely that proton decay will ever be witnessed; the experiments are rapidly approaching the theoretical limit of accuracy.

If proton decay does turn out to be a blind alley, more and more attention will be given to the only other experimental handle we know that can provide a glimpse of physics at the unification scale, and that is the magnetic monopole.

Magnetic monopoles

The legendary symmetry and beauty of Maxwell's electromagnetic equations have frequently been mentioned in earlier chapters. There is, however, a curious blemish that mars the otherwise perfect elegance of this theory. The equations are lopsided in their treatment of electricity and magnetism. Although these two forces are deeply interwoven, they do not enter into the theory entirely symmetrically. Electric fields are produced either by electric charges or by changing magnetic fields, while magnetic fields are produced by changing electric fields only. There seems to be no compelling reason, though, why magnetic fields cannot also be produced by magnetic charges (and electric fields by magnetic currents).

An ordinary bar magnet possesses a north pole and a south pole, but a deeper analysis reveals that the magnetism is actually produced by electric currents circulating at the atomic level. Because a loop of current inevitably produces a *pair* of magnetic poles, a north on one side of the loop and a south on the other, the magnet will be a 'dipole', i.e. have both a north pole and a south pole. It is no more possible for a loop of current to produce only one pole than it is for a coin to have only one face. Hence it is impossible to chop a single pole – a 'monopole' – out of a bar magnet.

Investigation shows that all magnets are dipoles. Magnetic monopoles, if they exist, must be exceedingly elusive. Systematic searches of rocks, including moon rocks and material from the ocean floor, have failed to reveal even a single pure magnetic charge. This led many physicists to suppose that magnetic monopoles do not exist. If so, magnetism is only ever a by-product of electricity. This would mean accepting that nature is unbalanced between electricity and magnetism.

In 1931 the British theoretical physicist Paul Dirac found that a place definitely exists in quantum physics for magnetic monopoles, even though nature may have chosen not to avail herself of this possibility. Dirac related the existence of magnetic monopoles to the phases of quantum waves, and in so doing found an intriguing connection between electric and magnetic charges. If a magnetic monopole did exist, claimed Dirac, the magnetic charge it carried would have to be a multiple of a fixed basic quantity, which in turn is determined by the fundamental unit of electric charge. Therefore if a monopole does show up, at least we know what quantity of magnetic charge to expect.

Although Dirac's analysis found a place in physics for magnetic monopoles it did not compel them to exist. For nearly half a century little more was written on the subject. Then, in 1975, the physics community was jolted by the announcement that a magnetic monopole had been discovered among cosmic rays. The claim turned out to be a false alarm, but it did give a boost to a rekindled interest in the topic. What led to this mood of excitement were some new theoretical ideas that took the monopole concept well beyond the work of Dirac. In essence, the theorists had discovered that magnetic monopoles are a more or less inevitable consequence of GUTs.

The grand unified monopole (or GUM) was invented by 't Hooft, and also Alexander Polyakov in Moscow. Their theoretical work suggested that, if GUMs exist, they will have some pretty odd properties. For a start, each monopole will possess something above the unification mass, i.e. nearly 10^{16} proton masses, making them as heavy as an amoeba. They would not be point particles. Instead, they would possess a complex internal structure consisting of zones of force rather like the layers of an onion.

As the direct production of such hugely massive particles is out of the question, monopole enthusiasts turned to cosmology. Perhaps GUMs were made along with ordinary matter in the big bang, and remain today as relics? Calculations were eagerly performed to see how many monopoles one might reasonably expect to be left over. To the theorists' acute embarrassment there seemed to be a superabundance of GUMs. Indeed, according to one estimate, magnetic monopoles should be as common as atoms in the universe. Clearly something was badly wrong. Stringent limits on the abundance of monopoles come from examining the magnetic field of the galaxy. Even at best, monopoles would be outnumbered by atoms by a factor of 10^{16}.

Theorists are still divided on what to make of this, and refer to the conflict as 'the monopole problem'. In Chapter Twelve we shall look at some recent ideas that seem to solve the monopole problem in a very neat way. Meanwhile, attention has turned to some of the likely consequences that could be expected if magnetic monopoles really are plying the universe with something like the maximum abundance allowed by astronomical observations. Estimates suggest that up to 200 monopoles a year could still be striking each square kilometre of the Earth's surface from space. If just one of these could be detected, it would provide spectacular confirmation of grand unification.

This prospect has stimulated a number of searches for cosmic monopoles using loops of electric current. The principle behind these experiments centres on the properties of certain materials called superconductors, which completely lose all resistance to electricity when they are cooled to a very low temperature. Superconductivity is essentially a quantum effect, and one of the important properties of an electric current flowing around a superconducting loop is that the magnetic field it generates is 'quantized', i.e. it comes only in certain fixed units of flux. Should a magnetic monopole happen to pass directly through the loop, then the flux will be seen to jump by a known number of quantum units.

On 14 February 1981 Blas Cabrera of Stanford University detected just such a flux jump. The observation caused something of a sensation, and was hailed by the experimenter as the first definite evidence for a magnetic monopole from space. Other groups rushed to set up their own experiments to confirm Cabrera's result, but so far without success. At the time of writing there is a growing belief that Cabrera's monopole might turn out to have been yet another false alarm.

Meanwhile, other theorists have been busy working out what else we might see if the Earth is being peppered by monopoles from space. One of the distinctive features of a GUM is its enormous mass, which implies a huge reserve of energy – 10^{16} times as much energy as can be released by a nucleus of uranium in a nuclear reactor. In fact it would take the energy content of only a few dozen monopoles per day to meet the power requirements of the average householder.

To release this energy you have to annihilate a monopole with an antimonopole, which means a north pole with a south pole. The creation of every north pole is accompanied by the creation of a south pole, and so on average there would be as many norths striking the Earth as souths. Because magnetic monopoles would be stable when immersed in ordinary matter they could in principle be systematically collected, segregated into norths and souths, and stored in some sort of electromagnetic 'bottles'. When the time came, a few norths could be mixed with a few souths to release astonishing quantities of energy. On a large scale such an arrangement could, of course, be used as an horrifically efficient binary weapon.

Some geophysicists have speculated that something of this sort might be occurring naturally inside the Earth. The incoming monopoles are slowed up as they plough through the Earth's interior, whereupon they sink towards the core and accumulate. The effect of the geomagnetic field would be to draw the norths northward and the souths southward, effectively preventing mixing. At periods of geomagnetic reversal, however, the two populations would swap places, and during their migration in opposite directions many north–south encounters would occur, causing wholesale annihilation. It has even been suggested that this process might be largely responsible for the Earth's interior heat.

The searches for proton decay and magnetic monopoles represent a dwindling hope that physics at the unification scale might yet be glimpsed experimentally. It is too soon to write these searches off, but many physicists are coming to the conclusion that the initiative now lies with the theorists. Few theoretical physicists believe that the grand unified theories are the last word. After all, GUTs succeed in merging only three out of the four fundamental forces. What new vistas might become visible if a truly unified theory was available?

Superforce

Gravity is nature's odd man out. The other three forces of nature can all be represented by fields of forces extending through space and time, but gravity *is* space and time. Einstein's general theory of relativity describes gravity as a warp field, a field of curvature in the geometry of spacetime. It is nothing but distorted emptiness.

The geometrical nature of the gravitational field might be elegant, but it has serious consequences for any quantum description. In fact, for decades, Einstein's general theory of relativity has resisted all attempts at a consistent quantum formulation. In spite of the fact that it is a gauge field, the description of gravity in terms of graviton messenger exchange yields sensible answers to only the simplest sorts of processes. The difficulty, as always, lies with the infinite terms which arise whenever closed graviton loops occur.

The infinity problems of the gravitational field are exacerbated by the fact that the graviton is itself gravitationally 'charged'. In this respect it resembles the gluon, which is the mediator of the strong force, yet it also carries colour 'charge'. Because all forms of energy – including gravitons – are a source of gravity, we might say that gravitons gravitate. This means that two gravitons can interact by exchanging a third graviton, as shown in Figure 21. More elaborate networks of gravitons are readily envisaged, and it is clear that closed graviton loops (Figure 22) will rapidly proliferate once processes other than simple graviton exchange are entertained.

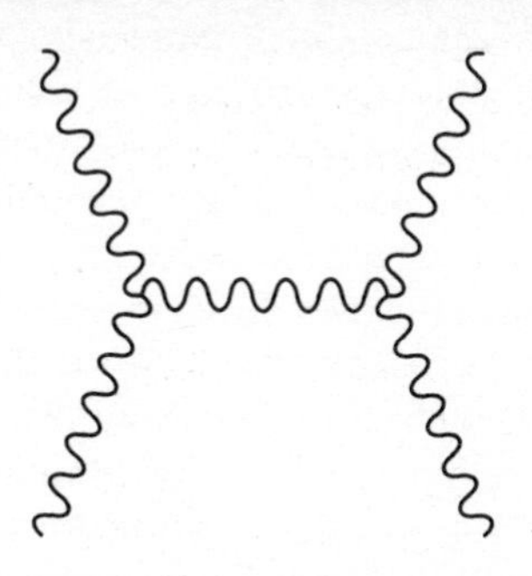

Figure 21. Gravity gravitates. Gravity is itself a source of gravity, an effect which at the quantum level permits gravitons (wavy lines) to interact with each other. In this process, two gravitons experience a mutual gravitational force by exchanging a third, messenger graviton.

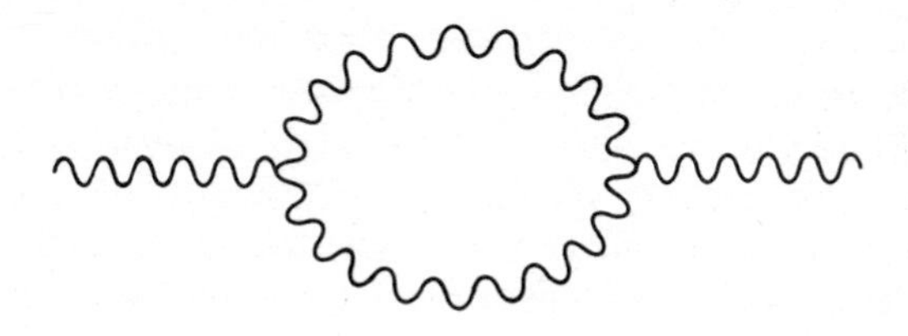

Figure 22. The ability of gravitons to interact with each other permits processes involving complex networks of graviton loops. A single loop is depicted here. These loops produce uncontrollable infinite terms in the mathematical description of quantum gravity, and render the theory essentially useless.

The appearance of endless strings of infinities in the graviton equations is now accepted as a deep-rooted malaise of quantum gravity based on Einstein's original theory. The situation is reminiscent of the weak force before it was united with electromagnetism. Both theories are 'non-renormalizable'. In the case of the weak force the answer lay with symmetry. The old theory did not have enough of it. Once a powerful gauge symmetry was built in, the infinities fell away like magic. Guided by this lesson, theorists began about ten years ago to search for a new symmetry, more powerful than any known before, which would rescue gravity from the poison of non-renormalizability. They came up with *supersymmetry*.

The essential idea of supersymmetry centres on the concept of spin, as it is understood in particle physics. When physicists talk of a particle having spin, they mean something rather more than the simple notion of a tiny ball rotating about an axis. Some of the weirdness of particle spin was discussed in Chapter 2. A particle with spin has a curious 'double image' view of the universe, something quite at odds with geometrical intuition.

The existence of spin is so fundamental to the nature of particles that it divides them into two distinct classes. In one class are the 'bosons', particles that either do not spin (i.e. have spin 0), such as the hypothetical Higgs particle, or have a whole number of units of spin. These include the photon, W, and Z, all with one unit of spin, and the graviton with two units. The behaviour of bosons is close to intuition as far as their spin is concerned.

In the other class are the 'fermions'. These have half-integral units of spin. All quarks and leptons are fermions, each having spin of one-half a unit. It is the fermion's spin which so taxes the imagination with its double-rotation property.

The distinction between bosons and fermions is a big one in physics. It shows up particularly strongly when these particles are assembled *en masse*. The fermions – the ones with half-integer spin values like electrons – are possessed of a sort of xenophobia, and do not permit their siblings to come too close. When quantum physics was in its infancy Pauli formulated his famous 'exclusion principle' which forbids any two electrons from sharing the same quantum state. It means that if you have a container of fixed size, there are only so many electrons you can put into it before they start to protest. (The effect has nothing to do with their electric repulsion, which is a separate matter. Neutrinos or neutrons are equally antisocial.) The Pauli principle was a notable early success for quantum physics because it explained why the electrons in heavy atoms don't all seek out the same lowest energy state and get into a tangle. Instead, they stack up in an orderly fashion, filling each energy level in turn. Without this discipline, chemistry would be a very chancy business indeed. Using Pauli's important rule the organization of the periodic table of chemical elements is immediately explained.

The Pauli principle explains a great deal more about the world, such as the difference between electric conductors and insulators. In a conductor some electrons are free to absorb energy from an applied electric field and accelerate into higher energy levels. In an insulator this cannot happen because all accessible higher levels are already blocked by other electrons. Another success for the Pauli principle is an explanation for the force which supports white dwarf stars and neutron stars against gravitational collapse. Without it, all stars would end their days as black holes.

In contrast to the isolationist behaviour of fermions, bosons positively love to get together. There is no objection to any number of them sharing the same bed, as it were. No exclusion principle operates here, and so the behaviour of massed bosons is completely different from their fermion cousins. They can be squashed and squeezed into the same state and the same space without protest. Because of this cooperative behaviour individual bosons can work together as a team, re-inforcing rather than frustrating each other's activities. In this way, vast numbers of bosons can act in orchestration and produce macroscopic effects that we can perceive directly. For example, armies of photons can merge coherently to build up a well-defined electromagnetic motion such as a radio wave. Fermions could never do this because they would get in each other's way. That is why, in spite of the fact that electrons also have an associated wave, we never see macroscopic electron waves.

The strong physical differences between fermions and bosons have led physicists, over the decades, to pigeon-hole them in rather separate regions of the brain. In particular, all the messenger particles are bosons, whereas quarks and leptons are all fermions. This means that bosons tend to be associated with *force*, whereas fermions are associated with *matter*. This sharp demarcation perhaps explains why, when supersymmetry was first proposed in the early 1970's many physicists were taken aback, for supersymmetry *unites* bosons and fermions into a single theory. It might seem like a shotgun marriage, the two groups being so distinct in their properties, but it can be accomplished by appealing to a still more powerful symmetry operation than the Lorentz–Poincaré symmetry, the symmetry which lies at the foundation of the theory of relativity. Mathematically, a supersymmetry operation resembles taking a square root of the Lorentz–Poincaré symmetry. Physically, it corresponds to changing a fermion into a boson or vice versa. Of course, we can't actually do this in the real world any more than we can twiddle a knob to reset the electron-ness of an electron in the gauge symmetries discussed in earlier chapters. Nevertheless, the operation can still be explored mathematically, and theories can be built which embody supersymmetry as a property.

It was not long before the supertheorists turned their attentions to gravity. Supersymmetry is closely related to geometry: if you carry out two supersymmetric operations in succession you get a simple geometrical operation like a shift in spatial position. In fact, supersymmetry mathematics has been called the square root of geometry. Gravity, being purely curved geometry, receives natural expression through the language of sypersymmetry, which brings out its gauge field nature in a more powerful way.

Because it unites bosons and fermions, sypersymmetry incorporates particles with different spins within the same family. Collections of particles, some with spin 0, others with spin ½, 1, and so on, can be grouped so that the family as a whole is supersymmetric. If one demands, therefore, that gravity is a supersymmetric theory it means that the graviton, with its spin value 2, cannot exist alone. It must belong to a whole family of particles which connect with spin 2 via the supersymmetry operation. These include particles of spin 0, ½, 1, and, significantly, 3⁄2. No elementary particle with spin 3⁄2 has ever been seen (though three quarks can align to give a net spin of 3⁄2), and so the predicted existence of such an entity is one of the novel features of supersymmetry.

Descriptions of gravity in these terms are known as the theory of *supergravity*. The way that supergravity differs from ordinary gravity is that the graviton is not the only particle given the responsibility of transmitting the gravitational force. This is done by a whole supersymmetric family, and particular assistance is due to the enigmatic spin 3⁄2 particles which physicists have dubbed 'gravitinos'.

The precise details of the family structure depend on how the theorist chooses to represent the supersymmetry mathematically. The most powerful representation is known as '$N = 8$' supergravity, and it generates a family of particles of impressive size: 70 with spin 0, 56 with spin ½, 28 with spin 1, and 8 with spin 3⁄2, as well as the lone graviton with spin 2, of course. An interesting question then arises. Can all these particles be identified with the known particles in nature, i.e. with the quarks, leptons and messengers? If so, we shall have at our disposal a unified theory of nature, which not only aggregates all particles of matter into a single superfamily, but also amalgamates all the messengers, and hence all the forces. Supergravity thus provides a framework for *total unification*, in which the entire world is placed under the control of a single, master force – a superforce – which displays itself through different facets – electromagnetism through photon messengers, the strong force through gluons, etc. – but all connected through supersymmetry (*see* Table 5).

In fact, supergravity goes beyond this. It provides a unified description of *force* and *matter*. Both force and matter originate with quantum particles, but the photons, Ws, Zs, and gluons are all bosons, whereas quarks and leptons are fermions. In supersymmetry these are united. Indeed, just as the graviton has gravitinos to go with it, so the other messengers team up with new particles called photinos, winos, zinos, and gluinos!

Table 5

Force				
Electricity	Electromagnetism (Maxwell, 1850s)	Electroweak force (Salam, Weinberg, 1967)	Grand unified theories (Glashow *et al.*, 1974)	Superforce (1990?)
Magnetism				
Weak force				
Strong force				
Gravity				

Successive unification of the forces of nature began with Maxwell's synthesis of electricity and magnetism in the nineteenth century. The union of the weak and electromagnetic forces is now well-established with the discovery in 1983 of the W and Z particles. Evidence for grand unification remains elusive, but eagerly sought. The theoretical foundations for a superunified theory which merges all the forces of nature into a single superforce are progressing rapidly.

The existence of all these 'inos' has a crucial effect on the mathematics of the theory, particularly in relation to the vexed question of renormalizability. Crudely speaking, the 'inos', which are fermions, generate infinities in the theory that have the opposite sign from the infinities caused by the bosons, such as gravitons. There is thus a tendency for cancellation, with gravitino-loop negative infinities cancelling graviton-loop positive infinities. In essence, the infinities supersymmetrize each other to death.

From the early days of supergravity there has been a single burning question: is supersymmetry a powerful enough symmetry to render supergravity renormalizable? Answering the question has not been easy. Supergravity is now a major industry, occupying the attentions of dozens of theorists and generating hundreds of research papers each year. The detailed mathematics has become so elaborate that few outside the immediate circle of cogniscenti have the slightest clue about what all the symbols mean. I have a supergravity expert in my department, and he will routinely accumulate a stack of paper 10 cm high as part of a single calculation. The point is that, however simple and elegant the mathematical foundations of a theory may be, checking the details can be a very messy business indeed.

Because of this tedium and complexity, no quick answers to the renormalizability question have been forthcoming, but those that have been attained are extremely encouraging. What seems to be happening is that supergravity is going beyond mere renormalizability, where infinites remain in the theory but are side-stepped by mathematical trickery. Instead, supergravity is clearly trying to find *finite* answers; in fact, in all calculations performed so far, the answers have without exception turned out to be finite. There is a strong belief that in supergravity the disastrous infinities that have plagued field theory for two generations have at last been eradicated.

Supergravity is the crowning achievement in the long search for unity in physics. Although still in its formative stages, it undoubtedly holds out great hope for solving three major outstanding problems of theoretical physics, i.e. how to unify all four forces of nature into a single superforce, how to explain the existence of all these fundamental particles – they are needed to maintain supersymmetry – and why gravity is so much weaker than the other forces of nature.

Confidence in the outcome of these investigations is so high in some quarters that Stephen Hawking foresees $N = 8$ supergravity as the culmination of theoretical physics. It could in principle explain everything

– all forces and all particles – in the physical world. If he is right – and it is perhaps too soon for theoretical physicists to start looking for jobs in biochemistry departments – then supergravity would differ in a fundamental respect from all other theories of physics. Until now, physical theories have been regarded as merely models which approximately describe the reality of nature. As the models improve, so the fit between theory and reality gets closer. Some physicists are now claiming that supergravity *is* the reality, that the model and the real world are in mathematically perfect accord. That is a very powerful philosophical position, and it is a measure of the euphoria generated by recent successes.

Set against this theoretical excitement are the very poor prospects for testing many of the new ideas by experiment. At the Royal Society meeting, Weinberg focussed on this impasse. 'Quantum gravitation seems inaccessible to any experiment we can devise,' he said. 'In fact, physics in general is moving into an era where the fundamental questions can no longer be illuminated by conceivable experiments. It's a very disquieting position to be in.' I asked him whether this meant that physics was degenerating into pure philosophy. 'I don't think so,' he replied. 'I think that the ingenuity of the experimentalists will find a way out of this.' But he had to admit he couldn't think of what that way might be.

At the time of writing, then, the unification of physics has leapt forward, and the outline of a complete theory of nature can at last be dimly perceived, even if experimental test seems remote. Like many compelling images it may turn out to be a mirage, but for the first time in the history of science we can form a conception of what a complete scientific theory of the world will look like.

10

Do We Live in Eleven Dimensions?

The first unified field theory

Much of the fascination of physics lies in the fact that it frequently explains the world in terms of things we don't see, and may not even be able to visualize however hard we strain our imagination. We have already encountered several examples, such as the spin of particles, wave-particle duality, and elastic space. Some people find this abstractness maddening or even dismaying; for others it is exhilarating and intriguing. Those who enjoy science fiction cannot fail to find the new physics a bonanza of weird ideas.

A classic example of the use of abstract concepts to explain nature occurred in 1915 when Einstein published his epoch-making general theory of relativity, one of those rare pieces of work that mark a turning point in mankind's perception of the world. The beauty of Einstein's theory rests not only in the power and elegance of his gravitational field equations, but in the sweeping radicalism of its conceptual basis, for Einstein not only swept away Newton's gravitation *and* his mechanics at a stroke, but abolished the very notion of gravity as a force. The general theory of relativity firmly established the idea of gravity as a field of geometrical distortion. Einstein thus reduced gravity to pure geometry. Where there used to be a pull across space, now there is a 'spacewarp'.

Einstein's theory was such a momentous advance in outlook that it was inevitable that the other forces of nature would soon come under renewed scrutiny. At the time, the only other force that had been definitively identified was electromagnetism. This force, however, did not appear to resemble gravity at all. Moreover, it had already received a very successful description by Maxwell several decades before, and there was no evidence that Maxwell's theory was to be doubted.

It was a lifelong dream of Einstein's that a unified field theory could be constructed in which all the forces of nature are merged into a single descriptive scheme based on pure geometry. Indeed, he devoted the greater part of his later years to the search for such a scheme. Ironically, the best hope we have for realizing Einstein's dream stems from the work of a little-known Polish physicist, Theordor Kaluza, who as early as 1921 laid the foundations for an astonishing new approach to the unification of physics, one which is breathtaking in its audacity.

Kaluza was inspired by the power of geometry to describe gravitation, and he wished to extend Einstein's work to include electromagnetism in the geometrical formulation of field theory. He had to accomplish this without altering Maxwell's sacrosanct electromagnetic equations. What he did provides a classic example of creative imagination and physical insight. Kaluza realized that there was no way that Maxwell's electromagnetic theory could be turned into geometry as we usually understand the word, even allowing for spacewarps. His resolution was brilliantly simple. He enlarged geometry just enough to accommodate Maxwell's theory. The way in which he did this is at once bizarre and unexpectedly persuasive. Kaluza showed that electromagnetism is actually a form of gravity, but not the gravity of familiar physics. It is the gravity of an unseen dimension of space.

Physicists have long been used to regarding time as the fourth dimension. The theory of relativity reveals that space and time are not by themselves physically universal qualities. Instead, it is necessary for them to be unified into a single four-dimensional structure, called spacetime. What Kaluza did was to go beyond this and postulate that there exists yet another dimension, an additional dimension of space, thus making a total of four space dimensions, or five dimensions in all. When this is done, Kaluza demonstrated, something of a mathematical miracle occurs. The gravitational field in this five-dimensional universe behaves just like ordinary gravity *plus* Maxwell's electromagnetic field, when viewed from the restricted perspective of four dimensions. What Kaluza was saying in his bold conjecture was that if we enlarge our vision of the universe to five dimensions there is really only one force field, and that is gravity. What we call electromagnetism is only that part of the gravitational field which operates in the fifth dimension, the extra dimension of space we have failed to recognize.

Kaluza's theory not only amalgamates gravity and electromagnetism into a single theory, it provides a geometrical formulation for both these force fields. An electromagnetic wave in this theory, such as a radio wave, is nothing other than a ripple in the fifth dimension. The distinctive motion of electrically charged particles in electric and magnetic fields is beautifully accounted for by supposing that they are cavorting about in this fifth dimension. Viewed this way, there are no forces at all, only warped five-dimensional geometry, with particles meandering freely through a landscape of structured nothingness.

The fact that, mathematically, Einstein's gravitational field in five dimensions is exactly and completely equivalent to gravity plus electromagnetism in four dimensions is surely more than a passing coincidence. Yet, as it stands, Kaluza's theory remains puzzling in one fundamental respect: we do not see the vital fourth dimension of space. The space of our perceptions is clearly and unalterably three-dimensional. If there is a fourth dimension of space, where is it? Before answering this question, it is as well to be completely clear what the concept of dimensionality actually means.

What are dimensions?

Science fiction writers have long extolled the virtues of extra dimensions of space. Often authors appeal to 'other dimensions' to get their characters from place to place in the universe without the tedium of traversing ordinary three-dimensional space at the comparative snail's pace of the speed of light or less. In Arthur C. Clarke's book *2001: A Space Odyssey*, an expedition to Saturn ends up plunging headlong through a gateway into another dimension located in one of Saturn's moons.

Mankind's fascination with dimensionality did not begin with science fiction, however. The early Greeks had a keen appreciation of its significance for the science of geometry. One curious example that brought the Greek geometers face to face with the puzzles of dimensionality concerned the properties of polygons (closed figures made up of sides of equal length, such as squares, pentagons, octagons, etc.). The number of polygons is unlimited; they may have any number of sides. By contrast, there are only *five* types of regular polyhedra (closed surfaces made up of regular shapes such as triangles, fitted together). As ever, the Greeks imbued their geometry with deep mystical significance, and Ptolemy even wrote a study on dimensionality in which he argued that no more than three dimensions of space are permitted in nature.

In the modern period, mathematicians such as Riemann developed a systematic study of higher-dimensional spaces for their intrinsic interest. A basic problem that they encountered, however, was the construction of a satisfactory definition of dimensionality. This was obviously important given that the mathematicians wished to prove rigorous theorems about the properties of spaces with different dimensions.

Intuitively, we divide up geometrical structures into one, two, or three dimensions according to the nature of their extension. Thus, a point, having no extension, is zero-dimensional. A line is one-dimensional, a surface two-dimensional and a volume three-dimensional. We can do a lot worse than recount the definitions given by Euclid himself, circa 300 B.C.:

> 'A point is that which has no part.
> A line is breadthless length.
> A surface is that which has length and breadth only.
> A solid is that which has length, breadth and depth.'

Euclid went on to point out that the extremities of a line are points, the boundary of a surface is a line, and the boundary of a solid is a surface. This led to the idea of defining dimensionality in a hierarchy, starting with zero for a point, and then building up one by one. Thus, a one-dimensional object is that which has points as its extremities, i.e. it is a line. In this way we arrive inductively at the definition of a four-dimensional structure as that which is bounded by a three-dimensional volume. There is no limit to the number of dimensions which can *logically* be defined this way, although the procedure has nothing to say about the real physical situation.

A more precise grasp of three-dimensionality can be obtained by imagining a scheme to label points in space. Suppose, for example, you wished to meet a friend at a pre-arranged spot. One method would be to give a longitude and latitude; you might choose the coordinates for the Empire State Building, for example. This still leaves freedom to specify altitude. Which floor will you be on? In all, three independent numbers are needed to uniquely define a point in space. For this reason space is said to be three-dimensional.

The theory of relativity revealed how space is interwoven with time, and that we should really think not of space alone, but of spacetime. What day are you going to meet your friend in the Empire State Building? To fix the time of an event requires a single number – the date – and so time is one-dimensional. Putting space and time together, then, we arrive at four-dimensional spacetime.

When we try to imagine extra dimensions, say a fourth space dimension, making five dimensions of spacetime in all, intuition fails. One way to get some help is to resort to analogy. Imagine a two-dimensional 'pancake' creature which is destined to spend its life for ever confined to a surface. It has no concept of 'up' or 'down'. Figure 23 depicts the pancake's universe. We can perceive that the surface is, in fact, enveloped in a three-dimensional space, but the pancake itself does not enjoy our broader perspective. It perceives only events that occur within the surface.

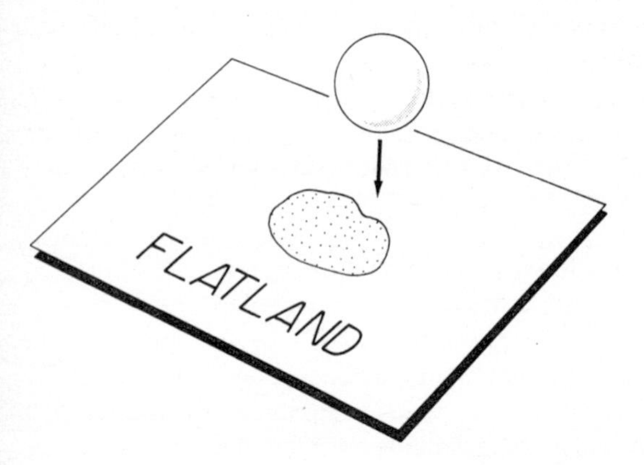

Figure 23. Two-dimensional universe. A pancake-type creature living in 'Flatland' has no perception of 'up' or 'down'. The ball about to penetrate his surface world will be perceived as a changing two-dimensional shape within the surface.

We can ask what the creature sees when a three-dimensional object penetrates the surface. The surface slices through a section of the object, a section which will generally change size and shape as the object passes through. A sphere, for example, would first be detected as a dot, which then swells to a disc, reaching a maximum radius, thereafter shrinking until it disappears at another dot. More complicated objects would produce more complicated shapes.

Reasoning by analogy we might suppose that the four dimensions of our spacetime are enfolded in a universe of five (or even more) dimensions, the geometry of which we cannot imagine, but which will nevertheless possess a perfectly logical description in terms of mathematics. Indeed, mathematicians long ago extended the rules of geometry to spaces with any number of dimensions (including infinity). It is therefore possible to make good sense of higher-dimensional spaces even though only three are apparent to our perceptions.

What sort of features would a four-dimensional space possess? One aspect of dimensionality concerns the number of mutually perpendicular directions that can be found. The surface of this page, for example, is two-dimensional. Place it flat on a table. The edges of the page at one corner define two lines at right angles. It is impossible to draw a third line from this corner anywhere across the page that is perpendicular to both edges. Such a direction can be found, however, if we are willing to go outside the plane of the page and draw a vertical line. Thus, in three-dimensional space, as opposed to the two-dimensional surface of the page, three mutually perpendicular directions exist.

In a four-dimensional space, it would be possible to find *four* mutually perpendicular directions. Figure 24 shows the situation for three

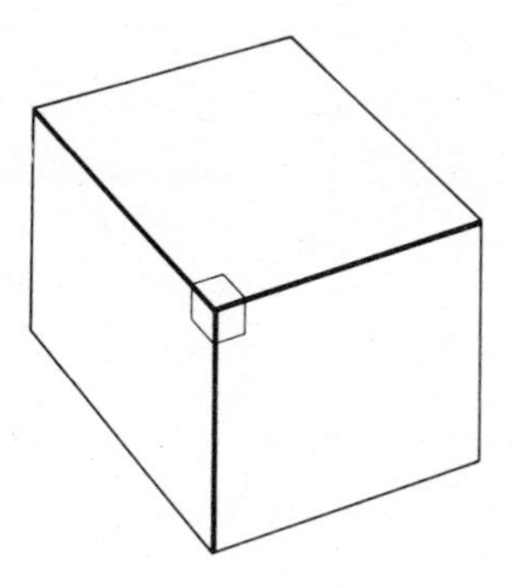

Figure 24. The edges of a rectangular block form three mutually perpendicular lines. No line in the three-dimensional space of our perceptions can be drawn from the corner of the block that is perpendicular to all three of these lines.

dimensions, with three mutually perpendicular lines making up the maximum number possible. Try as we may, we can never find a line at right angles to those three within the confines of ordinary space. Any line perpendicular to the three must go off in a direction that does not lie within our space at all. Although we cannot readily imagine *where* such a line goes, it is clear that *logically* such a line could exist. We could describe it. Its geometrical properties could be evaluated and catalogued.

A simple example of such a property is provided by a famous geometrical theorem learned by all high school children, due to the Greek geometer Pythagoras. The theorem concerns right-angled triangles. In Figure 25 the lengths of the sides of a right-angled triangle are a, b, and x. In symbols, Pythagoras' theorem states that these three lengths are related by a simple formula: $x^2 = a^2 + b^2$. To take a convenient case, if we know that $a = 3$ and $b = 4$, we deduce that $x = 5$, because $5^2 = 3^2 + 4^2$.

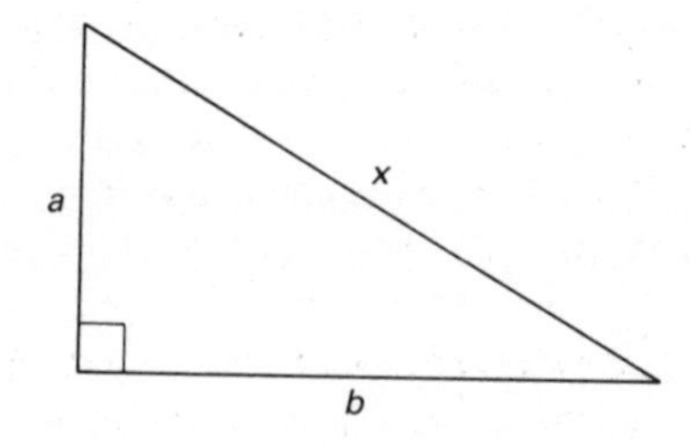

Figure 25. Pythagoras' famous theorem relates the lengths *a*, *b*, and *x* of a right-angled triangle. This theorem is readily generalized to higher dimensions.

The triangle shown in Figure 25 is, of course, a two-dimensional object, but we can readily generalize Pythagoras' theorem to three dimensions. In Figure 26 a rectangular box is depicted. The lengths of the sides are *a*, *b*, and *c*. The theorem now refers to the diagonal distance between opposite

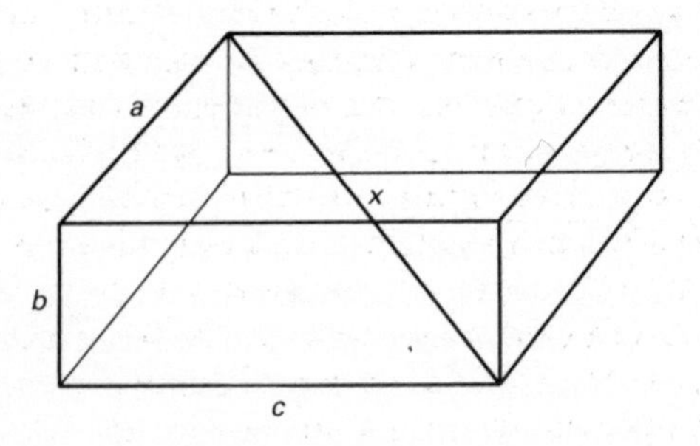

Figure 26. The length of the diagonal through the rectangular block can be related to the lengths of the block's sides, *a*, *b*, and *c*, by a simple generalization of Pythagoras' theorem. It is easy to extend this generalization to four or more space dimensions.

corners of the box, labelled x in the figure. In symbols, Pythagoras' theorem now states that $x^2 = a^2 + b^2 + c^2$. The form of this equation is closely similar to the two-dimensional case given above, but we now need the lengths of *three* mutually perpendicular sides, a, b, and c, to compute the length of the diagonal x.

In a four-dimensional space, the diagonal length would be computed from *four* perpendicular lengths, say a, b, c, and d. We should then have the formula $x^2 = a^2 + b^2 + c^2 + d^2$. So although we cannot imagine a four-dimensional box, we can still discuss the details of its geometrical properties.

Valuable though these geometrical insights are, they turn out to be a house of cards. The house collapsed when the modern era of mathematics began towards the end of the nineteenth century with the development of a powerful branch of mathematics known as set theory. Among the shocks which mathematicians suffered was the discovery by Georg Cantor that there are as many points in a line as there are in a surface. The intuitive idea that a surface is somehow infinitely richer in points than a single line drawn in that surface was utterly discredited. This bombshell left even respectable mathematicians incredulous. Some dismissed Cantor as mad. Charles Hermite wrote dismissively, 'To read Cantor's writings seems a veritable torture . . . The mapping between a line and a surface leaves us absolutely indifferent . . . such arbitrariness . . . the author would have been better inspired to wait . . .' etc.

Only after the turn of the century was the damage repaired and a satisfactory definition of dimensionality achieved. Several important contributions by L. E. J. Brouwer, René Lesbesgue, and others eventually established a watertight procedure for comparing two spaces and deciding whether they had the same dimensionality. These proofs are based on subtle and abstract ideas of set theory that are far removed from intuition. Only with such care and attention to detail are the logical foundations of our science, and our experience, made secure.

Why three?

Whatever the dimensionality of space really is, there is no doubt that only three dimensions are directly apparent to our senses. Many scientists have wondered if it is possible to understand why three has been selected by nature. Is the number in some sense unique?

In 1917 the physicist Paul Ehrenfest wrote a paper entitled 'In what way does it become manifest in the fundamental laws of physics that space has three dimensions?' Ehrenfest focussed his attention on the existence of stable orbits, of the type followed by the planets around the sun or by an electron around an atomic nucleus. The widespread occurrence of inverse square laws of force is well known in physics. In Chapter 5 we saw how gravitational, electric, and magnetic forces all satisfy such a law. As early as 1747 Immanuel Kant recognized the deep connection between this law and the three-dimensionality of space. The equations describing the gravitational or electric fields of a point source can readily be generalized to other dimensions, and solved. The solutions reveal that in a space with n dimensions we have to deal with an inverse $n - 1$ power law. Thus, in three dimensions $n - 1 = 2$ and the law is inverse square. In four dimensions $n - 1 = 3$, so we obtain an inverse cube law. And so on. It is readily shown that if, say, the sun generated an inverse cube gravity field the planets would rapidly spiral into it and be engulfed.

The situation with atoms is similar. Even taking into account quantum effects, it turns out that electrons have no stable orbits in spaces with greater than three dimensions. Without stable atomic orbits, chemistry, and hence life, would be impossible.

Another phenomenon that depends sensitively on dimensionality is wave propagation. It is easily shown that in spaces with even numbers of dimensions a wave will not propagate cleanly, but will trail disturbances behind it that cause reverberation effects. For this reason it is impossible to transmit well-defined signals in a two-dimensional surface such as a sheet of rubber. Discussing this topic in 1955, the mathematician G. J. Whitrow concluded that advanced life would be impossible in a space of even dimensionality because living organisms require the efficient transmission and processing of information in order to operate coherently.

These studies prove not that other dimensions of space are impossible, only that physics in a world with other than three dimensions would be profoundly different, and possibly far less ordered than in the world we perceive.

How are we to square all this with Kaluza's theory of a universe with four space dimensions? One possibility is to regard the extra, unseen dimension as purely a device, a mathematical trick with no physical significance. A more appealing idea, however, was proposed not long after Kaluza published his original theory.

The theory of Kaluza and Klein

In 1926 the Swedish physicist Oscar Klein came up with a marvellously simple answer to the question of where Kaluza's fifth dimension has gone. Klein proposed that we do not notice the extra dimension because it is, in a sense, 'rolled up' to a very small size. The situation can be compared to a hose-pipe. Viewed from afar, the pipe appears to be just a wiggly line. On closer inspection, what we took to be a point on the line turns out to be a circle around the circumference of the tube (Figure 27). Suppose, conjectured Klein, that our universe is like that. What we normally think of as a point in three-dimensional space is in reality a tiny circle going round the fourth space dimension. From every point in space a little loop goes off

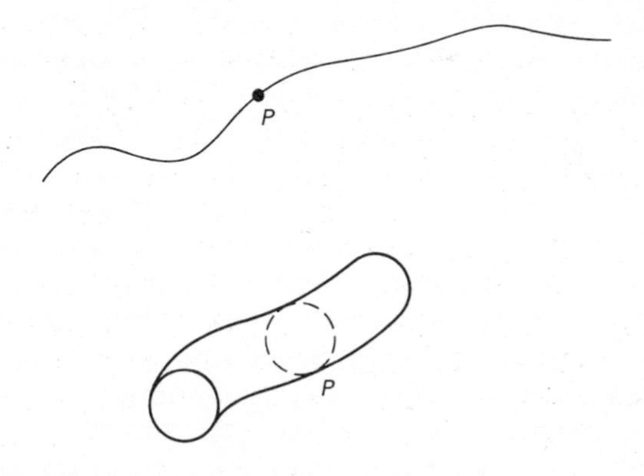

Figure 27. From a distance a hose-pipe looks like a wiggly line. On closer inspection a point P on the line turns out to be a circle around the circumference of the pipe. It is possible that what we normally regard as a point in three-dimensional space is really a tiny circle around another dimension of space. This idea forms the basis of Kaluza and Klein's unified theory of electromagnetic and gravitational forces.

in a direction that is not up, down, or sideways, or anywhere else in the space of our senses. The reason we haven't noticed all these loops is because they are incredibly small in circumference.

Klein's idea takes a bit of getting used to. Part of the trouble is that we cannot imagine *where* those loops are looping. The loops are not *inside* space, they extend space, just as a wiggly line moved rigidly on a loop marks out a tube. We can readily envisage the situation in two dimensions, but not in four. Nevertheless, the proposition still makes sense. There are no problems with stable orbits or wave propagation either, because matter and waves are not free to move off in an unbounded way in the extra dimension. The fifth dimension can still be *there*, but nothing can get very far by going through it. Of course, there is then no possibility of using the Kaluza–Klein theory to short-cut space travel – alas for fiction writers!

Klein was able to calculate the circumference of the loops around the fifth dimension from the known values of the unit of electric charge carried by electrons and other particles, and the strength of the gravitational forces between particles. The value came out to be 10^{-32} cm, or about 10^{-20} of the size of an atomic nucleus. It is no surprise that we have not noticed the fifth dimension, for it is rolled up far smaller than any structure yet discerned, even in subnuclear particle physics. There is clearly no question of, say, an atom going off round the fifth dimension. Rather, we should think of the extra dimension as somehow inside the atom.

In spite of its ingenuity, the Kaluza–Klein theory remained little more than a mathematical curiosity for over fifty years. With the discovery of the weak and strong forces in the 1930s, the idea of unifying gravity and electromagnetism lost much of its appeal. Any successful unified field theory would have to accommodate not just two, but four forces. This step could not be accomplished, therefore, until a proper understanding of the weak and strong forces was attained. In the late 1970s, with GUTs and supergravity fresh on theorists' minds, someone remembered the old Kaluza–Klein theory. It was promptly wheeled out of mothballs, dusted off, and remodelled to accommodate the now extended range of known forces.

In GUTs, the theorists' ability to subsume three very different sorts of forces under a single conceptual umbrella hinged, as we saw in earlier chapters, on the discovery that all three forces involved can be described in terms of gauge fields. The central property of these gauge fields is the presence of certain abstract symmetries, which is how they attain their power and elegance. The presence of symmetries in the force fields already hints strongly at some sort of hidden geometry at work. In the revitalized Kaluza–Klein theory the gauge field symmetries become concrete; they are the geometrical symmetries associated with the extra space dimensions.

As in the original version of the theory, the forces are accommodated by grafting more dimensions of space on to known spacetime, but the fact that we now have three forces to accommodate demands several additional dimensions. A simple count of the number of symmetry operations embodied in the grand unified force leads to a theory in which there must be *seven* extra dimensions, making ten space dimensions in all, plus time, or *eleven* spacetime dimensions. The modern version of the Kaluza–Klein theory postulates an eleven-dimensional universe.

Once again, it is necessary to suppose that the extra seven space dimensions are somehow 'rolled up' very small, so that we do not notice them. There is only one way to roll up one extra dimension, and that is into a circle. Higher dimensional spaces, however, can be compacted in many different ways. For example, a two-dimensional surface could be joined to form either the surface of a sphere or a doughnut (known as a torus). Though both structures are closed, and could be made very small, they differ profoundly in their topology; the doughnut has a hole through it.

When it comes to seven dimensions the range of topologies is enormous. Which shape is correct? One particularly attractive choice is the seven-dimensional analogue of the sphere, known simply as the seven-sphere. If the unseen space dimensions really do have this form it means that every point of three-dimensional space is in fact a minute seven-dimensional 'hyperball'. The seven-sphere attracted the attention of mathematicians over half a century ago because it possesses a number of unique features that endow it with rich geometrical properties. The details need not concern us here, but if nature had to find a closed geometrical structure that permits anything like the fundamental forces and fields that we perceive in the real world, then the seven-sphere would be the simplest choice. You couldn't get the sort of structures we see, from atoms to galaxies, out of some simpler mathematical arrangement.

A sphere is a highly symmetrical shape, and a seven-sphere contains many additional symmetries not found in an ordinary sphere. These are intended to model the underlying gauge symmetries of the force fields. However, one of the reasons it took so long for physicists to verify the forces is that symmetries are sometimes hidden, or broken, in the fashion described in Chapter 8. In the Kaluza–Klein theory this symmetry-breaking is achieved by distorting the shape of the seven-dimensional structure somewhat away from exact sphericity. The 'squashed' seven-sphere is thus emerging as the most favoured shape for the unseen compactified extra dimensions.

The resurrected Kaluza–Klein theory has proved so inspiring that physicists have been rushing to recast the laws of physics in eleven dimensions. One problem which arises is to understand why spacetime should adopt a seven–four split. Is it inevitable that seven out of the eleven dimensions should curl up to invisibility, leaving behind the four dimensions of direct experience, or is a different configuration possible, say eight–three?

In their search for a reason why seven dimensions should spontaneously compact themselves, theorists have been working on the assumption that physical systems always tend to seek out their lowest energy state. We met a good example of this principle in Chapter 8 with the ball on the 'Mexican hat' surface, where the ball eventually finds a stable state of lowest energy in the 'rim of the hat'. This suggests that a shrunken squashed seven-sphere is in some sense the lowest energy configuration of spacetime.

Alternatively, it is conceivable that the seven-sphere configuration is only one of many possible arrangements. It is fascinating to conjecture that far out in the cosmos, beyond the bounds of our observable universe, space possesses a different number of dimensions. Perhaps if we could travel thousands of millions of light years we should suddenly find ourselves in a universe of five space dimensions instead of three. If so, we may have an answer to the question, 'Why three?' It could be that eleven-dimensional spacetime organizes itself into domains of different apparent dimensionality. Because the structure of the force fields depends on the geometrical symmetries of the compactified dimensions, these forces would differ from one domain to the other. Added to these differences will be the many problems about stable orbits, wave motion, etc., discussed earlier. All this will ensure that the physical conditions in the domains that do not enjoy a seven–four split will be wildly dissimilar from those in our own cosmos. It is doubtful if life could flourish, or even exist at all, in these other domains. Living organisms are highly delicate, and seem to depend critically on the uniquely felicitous blend of forces found in our region of the universe. This suggests that we, as observers, have by our very presence selected a region of spacetime with three perceptible space dimensions. We simply could not live in any of the other differently dimensional domains which may exist.

Why eleven?

The use of 'anthropic' reasoning, as it is called, to explain why space appears to us as three-dimensional, prompts a further intriguing question. Must the number of overall spacetime dimensions inevitably be eleven, or might this too vary from place to place? One could envisage, say, a twenty-one-dimensional universe in which seventeen dimensions roll up to a compact structure, endowing the world with a far more complex pattern of force fields than the four of our experience. Who knows what intricate structures, what elaborate life forms might thrive in such a universe?

Throughout history, men and women have been fascinated by numerology. The early Greeks imbued certain numbers with mystical and metaphysical significance. To this day the integer four – the number of sides of a square – retains a vestige of its ancient associations with honesty and fairness in the expression 'a square deal'. Many people still have 'lucky' or 'unlucky' numbers such as three, seven or thirteen. The Bible makes repeated use of the numbers seven and forty. People still associate the number 666 with the devil.

When numbers occur naturally in the world, it is tempting to search for meaning behind them. Sometimes they seem to be purely accidental, as with the number of planets in the solar system. Other naturally occurring numbers suggest a deeper significance. The number of hadrons turned out to be a consequence of the number of available permutations of quark combinations. Is the dimensionality of spacetime merely an accident, just one of those things like the numbers of planets? Or is it a pointer to a profound truth about the logical and mathematical structure of the physical world?

There is a curious piece of evidence that the number eleven does, indeed, have a deep mathematical significance, and it comes from a branch of physics that, superficially at least, is quite unconnected with the Kaluza–Klein theory, namely supergravity.

In the previous chapter the most hopeful formulation of supergravity was discussed, known as $N = 8$. The cryptic designation '$N = 8$' needs some explanation. The supersymmetry operation connects together particles with different spin into a superfamily of 163 particles. It might be wondered why there are only 163 in the superfamily. If the supersymmetry operation shifts from a particle with one value of spin to a particle with another, why can't we go on shifting for ever and generate an endless sequence of particles with arbitrarily large spin? The answer is that, for supersymmetry to work properly as a symmetry, it has to involve a 'closed' sequence of operations. It can generate only a finite family of particles. As there are good mathematical reasons why particles with spin greater than 2 cannot exist, a superfamily with 163 members is the largest that can be entertained.

The designation $N = 8$ refers to the number of steps that connect particles with different spin across the whole range of available spins under the supersymmetry operation. Because spin can point both 'up' and 'down', its projection can vary from +2 (a spin-2 particle pointing up) to −2 (a spin-2 particle pointing down) in half-integer steps. There are eight such steps between −2 and +2, hence eight supersymmetry operations are needed to generate all the spin projections required to build the superfamily of particles. It also turns out that this number is related to the number of species of gravitinos, i.e. eight in this theory.

As usually formulated, the concept of spin refers to the rotational properties of a particle in good old three-dimensional space. Years ago, however, mathematicians amused themselves by inventing descriptions of spin in spaces of other dimensions just to see what things would look like. It so happens that, as far as supergravity is concerned, the theory appears a lot simpler if there are more than three dimensions available. In fact, the simplest description of all is with the theory recast in *eleven* dimensions. In eleven dimensions the eight separate symmetry operations of $N = 8$ supergravity collapse to just one – we get '$N = 1$' supergravity.

Imagine an enthusiastic mathematician who had no knowledge of the dimensionality of the real universe, but who on grounds of elegance and unity discovered supergravity. He would be drawn to formulate an eleven-dimensional theory of spacetime and conclude that, if nature knew what she was about, eleven would be the dimensionality of the real universe. Is this concurrence of the number eleven just an accident, or does it point to a deep connection between supergravity and the Kaluza–Klein theory? Many physicists hope that this connection really exists, and that the two traditions of unifying physics – supergravity and grand unified theories – will come together in a common description. Salam has written the following words:

> 'If this theory is correct, we may be very near to a final, complete unification of all forces with spinning matter, and with the fundamental charges being manifestations of hidden dimensions of space.'

The geometrization of nature

We have seen how Einstein's dream of a unified field theory built out of geometry has come very close indeed to realization. In the modern Kaluza–Klein theory all the forces of nature, not merely gravity, are treated as manifestations of spacetime structure. What we normally call gravity is a warp in the four spacetime dimensions of our perceptions, while the other forces are reduced to higher-dimensional spacewarps. All the forces of nature are revealed as nothing more than hidden geometry at work.

As long ago as 1870, the mathematician W. K. Clifford addressed the presitigious Cambridge Philosophical Society 'On the space theory of matter'. Clifford declared:

> 'That small portions of space are analogous to hills on a surface which is on the average flat . . . That this property of being curved or distorted is continually being passed on from one portion of space to another after the manner of a wave. That this variation of the curvature of space is what really happens in that phenomenon we call the *motion of matter*. That in the physical world nothing else takes place but this variation.'

These thoughts are remarkably prophetic of the general theory of relativity, developed nearly half a century later by Einstein. Clifford, however, seems to go beyond general relativity, and conjecture that, as well as forces, particles of matter are themselves nothing but bumps or kinks in empty space.

There is a deep compulsion to believe in the idea that the entire universe, including all the apparently concrete matter that assails our senses, is in reality only a frolic of convoluted nothingness, that in the end the world will turn out to be a sculpture of pure emptiness, a self-organized void. Geometry was the midwife of science. The painstaking work of generations of astronomers charting the paths of the heavenly bodies across the sky led eventually to the Newtonian revolution, and the explanation of the celestial patterns in terms of forces and fields. Now we have come full circle, and the forces and fields are themselves being explained in terms of geometry.

In the early 1960s the American theoretical physicist John Wheeler extended the work of Clifford and Einstein, and attempted to build a complete theory of the world based on the geometry of empty spacetime alone. He called this programme 'geometrodynamics'. It has as its aim the explanation of both particles and forces in terms of geometrical structures.

Wheeler's model for electric charge provides a good illustration of the general philosophy behind the project. He speculated that a charged particle is really a sort of entrance or gateway to a little tunnel that connects one point of space to another, rather like a miniature space bridge through another dimension. The remote end of the tunnel would be seen by us as another particle with opposite electric charge. Thus, the two ends of the 'Wheeler-wormhole' could be an electron–positron pair, for example. Whereas nineteenth century physicists would have said that the electric 'lines of force' focus and terminate on the charged particle, Wheeler proposed that the lines simply concentrate along it, to emerge intact at the other end (*see* Figure 28). In this way there need be no sources of electricity at all, only holes in space, trapping electric fields within them!

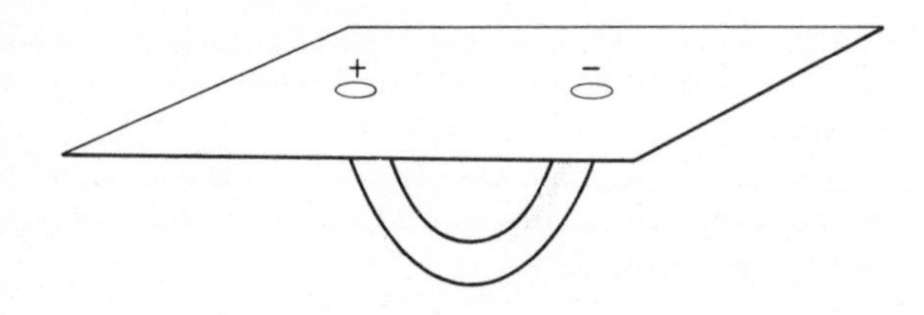

Figure 28. Wheeler conjectured that an electrically charged particle might be the end of a tiny tube or 'wormhole' passing through another dimension to link up again with our own three-dimensional space at the location of an oppositely charged particle somewhere else in the universe.

Geometrodynamics has many such delightful features, but it was never a complete success. Wheeler himself wrote that 'the most evident shortcoming . . . is that it fails to supply any completely natural place for spin ½ in general and the neutrino in particular'. In more recent years he has adopted the position that any theory which already assumes spacetime cannot also explain spacetime. In particular, the dimensionality of spacetime is built into the theory at the outset, and so it cannot emerge as a consequence of the theory. Any complete theory of nature has to account for the existence of the 'raw material' – spacetime itself – from which the geometrodynamical world is built. Wheeler believes that this can come only from a study of quantum physics, and he looks forward to a time when we will understand how the quantum, rather than spacetime, is the fundamental building block of reality.

With the benefit of hindsight we can see that the failure of Wheeler's geometrodynamics was due in part to its restriction to four dimensions. With the full range of eleven dimensions available, the variety and complexity of physical structures that could be built is greatly extended. In the Kaluza–Klein theory, particles are not treated as 'wormholes' in space, but as *excitations* within an eleven-dimensional geometry. The hope now is that this geometry will itself be explained, as Wheeler would wish, in terms of the quantum phenomenon.

Probing the hidden dimensions

Nature may be beautiful, but beauty alone will rarely convince physicists that a theory is correct. Hard physical evidence is also demanded. The power and elegance of the eleven-dimensional Kaluza–Klein theory obliges us to take it seriously, but if there is no conceivable way to verify that seven extra dimensions of space really exist, the theory loses much of its appeal.

Fortunately, however, it may be possible to demonstrate the existence of the other dimensions physically. For the theory to work, the seven extra space dimensions must be 'rolled up', probably in the form of a seven-sphere, to a circumference of 10^{-32} cm. To probe structures on this ultramicroscopic scale presents a major challenge: we do not have direct control over any objects this small, and so we cannot send something 'into' the seven-sphere to explore.

Quantum physics associates a scale of energy (equivalently mass) with a scale of length. Measure-for-measure the diameter of a nucleus (about 10^{-12} cm) corresponds roughly to the mass of the pion. As smaller lengths are probed, so the energies involved start to climb. To explore the quark-filled interior of a proton it is necessary to go up to energies at least ten times greater than the proton mass. Much further up the scale of energy is the unification mass, at about 10^{14} proton masses. If one had command of such huge mass-energy (which we don't) it would be possible to probe the world of the X particles, at which the very distinction between quarks and leptons fades away.

How much energy do we need to 'get inside' the seven-sphere and explore the other dimensions of space? According to the Kaluza–Klein theory it is necessary to go beyond even the unification scale, to an energy equivalent to 10^{19} proton masses. Only at this unimaginable energy would the extra dimensions manifest themselves directly.

The huge value of 10^{19} proton masses is known as the 'Planck scale', because it was originally discovered by Max Planck, the inventor of the quantum theory. At the Planck energy, all four forces of nature would be completely merged into a single superforce, and all ten dimensions of space would exist on an equal footing. If we could concentrate enough energy to take us up to the Planck scale, the full dimensionality of space would be exposed in all its splendour.

Letting imagination have free rein, it is possible to envisage mankind one day gaining control over the superforce. To achieve this would enable us to manipulate the greatest power in the universe, for the superforce is ultimately responsible for generating all forces and all physical structures. It is the fountain-head of all existence. With the superforce unleashed, we could change the structure of space and time, tie our own knots in nothingness, and build matter to order. Controlling the superforce would enable us to construct and transmute particles at will, thus generating exotic forms of matter. We might even be able to manipulate the dimensionality of space itself, creating bizarre artificial worlds with unimaginable properties. Truly we should be lords of the universe.

But how is this control to be gained? First we need to harness enough energy. To get some idea of what is involved, the Stanford linear accelerator is 3 km long and can boost electrons to the energy-equivalent of about twenty proton masses. To attain the Planck energy would require an accelerator some 10^{18} times longer, making it as long as the Milky Way – 100 000 light-years. Not the sort of project to be completed overnight!

In the unified theory of the forces we can distinguish three crucial thresholds, or energy scales. First there is the Weinberg–Salam energy, at about ninety proton masses, beyond which the electromagnetic and weak forces merge into a single electroweak force. The second is the unification energy at 10^{14} proton masses, marking the onset of GUT physics. Finally, there is the Planck energy at 10^{19} proton masses, representing the ultimate scale of energy at which all of physics comes together in spectacular simplicity. One of the great unsolved problems is to explain these three numbers, and in particular provide a reason why the gap between the first and second numbers is so enormous.

Present technology can take us only as far as the first of these thresholds. As we saw in the previous chapter, proton decay could provide an indirect means of probing physics at the unification scale, but there seems to be no hope at all of directly achieving the unification energy, let alone the Planck energy.

Does this mean we will never be able to see the raw superforce at work and perceive the unseen seven dimensions of space? It is certainly true that, using technology, we are fast approaching the end of the energy road with the proposed Desertron. Nevertheless, human technology does not exhaust the full range of physical circumstances. There is nature itself. The universe is a huge natural laboratory, and 18 000 million years ago the biggest particle physics experiment of all was carried out. We call it the big bang. We shall see that this primeval violence was enough to unleash the superforce, albeit momentarily, but perhaps long enough for it to have left a permanent trace of its erstwhile existence.

II
Cosmic Fossils

The origin of the elements

One sunny, spring day in 1822 a young country doctor, Gideon Mantell, was visiting a patient near his home town of Lewes in Sussex, England. Accompanying Dr Mantell for the ride was his wife Mary Ann, who took the opportunity to stroll along a country lane while her husband attended to his patient. Coming across a pile of quarried stones intended for road repairs, Mrs Mantell happened to notice a curious brown, shiny object. On closer examination it turned out to be a piece of sandstone containing several very large teeth. Mrs Mantell showed the teeth to her husband, an amateur geologist, who became very excited. The teeth were reminiscent of those of the iguana lizard, and Dr Mantell made the bold conjecture that the teeth his wife had chanced upon had once belonged to a species of huge plant-eating reptile that inhabited the Earth before the mammals existed. He call the creature *Iguanadon*. The Mantells had discovered and correctly identified the first dinosaur fossil.

Mrs Mantell's accidental discovery came at a critical time for science. By tradition, the age of the Earth was thought to be several thousand years, a belief re-inforced by the biblical account of Genesis. However, by the end of the eighteenth century, geology was becoming a properly scientific subject, and most geologists had begun to recognize that vast intervals of time were necessary for the completion of geological activity such as sedimentation and erosion. Still, even as late as 1779, the French geologist George Louis Leclerc put the age of the Earth at only 75 000 years. By the middle of the nineteenth century that time-span had been stretched to hundreds of millions, or even thousands of millions of years. Today, the Earth has been radioactively dated at 4 600 million years.

The new dinosaur fossils soon came to be recognized as the remains of now extinct creatures that roamed the Earth between about 200 and 65 million years ago. It is a remarkable thought that by examining today's rocks; we can deduce something about the world in such remote times past, 200 million years is such an enormous length of time that it defies human imagination. Subsequent, more diligent searches have uncovered fossil remains of living organisms that date from at least 3 000 million years and possibly nearly 4 000 million years.

Although most people associate fossils with the frozen imprints of once-living creatures, there are many other physical objects that are imprinted with an inanimate record of the remote past. For example, the pock-marked surfaces of the moon, Mars, and Mercury bear witness to a phase of violent bombardment at the dawn of the solar system. In a sense, all physical things are fossils. Every object that exists possesses some sort of history and encodes information about the circumstances that brought it into being. The trick is to be smart enough to decode the information.

An interesting game is to take the object that is most familiar to us – our bodies – and ask what they can tell us about the past.

For a start there is biological information. This is encoded in our genes which are built from the distinctively shaped molecules of DNA. All life on Earth is based on DNA which may therefore be regarded as a relic of the origin of Earth life, some 4 000 million years ago. Our particular genetic structure bears countless imprints of the physical conditions encountered by our ancestors over the aeons, and which helped to shape the evolutionary pathway followed by our species. Our bodies are thus living fossils that embody a coded history of our planet.

Biological information relates to the way in which the atoms of carbon, hydrogen, oxygen, and other elements within living organisms are strung together into complex forms. But what about the atoms themselves, the raw material from which our bodies and all objects around us are made?

According to modern ideas of cosmology, these atoms did not always exist, but are relics of physical processes that occurred long ago, in Earth's prehistory, out in the depths of the universe. They are cosmic fossils. As we saw in Chapter 2, the primary constituent of cosmic material is hydrogen, with helium accounting for about 10 per cent of all atoms, and the ninety or so other elements representing only a minute fraction of the whole. Much of the material inside us, then, consists of cosmic trace elements, hugely concentrated. Their origin must be sought in the complex processes that occur inside stars.

When the universe began, the cosmic material contained essentially no medium or heavy nuclei. These elements are the ashes of the nuclear fires that sustain the stars. In a star like the sun the core is a nuclear fusion reactor, in which the fuel consists primarily of hydrogen nuclei (protons). The intense heat of the solar furnace agitates the protons with such violence that occasional very close encounters occur, even though protons will repel each other with a powerful electric force. Should the colliding protons come within range of the strong nuclear force, fusion is possible. A nucleus consisting of two protons is unstable, but if one proton transmutes into a neutron through a weak interaction (essentially the inverse of beta decay) then a stable nucleus of deuterium forms, with the release of energy, which helps keep the furnace hot. Further fusion reactions result in the conversion of deuterium into helium. In old stars the synthesis of heavier nuclei out of light ones is more developed. Successive fusion processes produce first carbon and then a whole sequence of ever-more complex nuclei.

As a star approaches the limit of its fuel reserves, its internal structure resembles the skins of an onion – to use the analogy once more – with layer upon layer of different chemical elements representing various stages in the long sequence of synthesis. Over its lifetime an old star has gradually been transformed from almost pure primordial hydrogen and helium into a repository for spent nuclear ash, in the form of heavy chemical elements. During its final phase of activity, such a star may become unstable. The faltering nuclear reactions are unable to sustain the intense internal heat and pressure needed to support the star against its own tremendous weight. Gravity then runs out of control, imploding the star's core in an instant. A huge pulse of energy in the form of neutrinos and shock waves released from the core blasts the outer layers of the star into space, spewing the heavy elements into the depths of the galaxy. This outburst is called a supernova explosion (*see* Chapter 5). Each one enriches the galactic material with the trace elements so vital in the formation of solid planets like the Earth and the life-forms that inhabit it. Our bodies are therefore built from the fossilized debris of once-bright stars that annihilated themselves aeons before the Earth or sun existed.

The heavy elements in the world about us record a violent history, but the light elements, hydrogen and helium, date from a still more violent epoch in our cosmic history – the big bang. The question arises of whether these elements were simply there from the beginning or whether they are also fossils from some very early phase.

The intense heat which accompanied the big bang provides the key to an understanding of the early universe. In its simplest form, the hot big bang theory assumes that the universe exploded spontaneously into existence from a state of infinite compression and infinite heat. As the expansion proceeded, so the temperature fell from infinity, rapidly at first, and then more slowly, until the universe had cooled enough for stars and galaxies to form. Before about 100 000 years, the temperature remained above several thousand degrees, preventing the formation of atoms. For 100 millenia the cosmic material remained in the form of a glowing plasma of ionized hydrogen and helium. Only when the universe had cooled to something like the temperature of the sun's surface did the first atoms form. Atoms, then, are relics from 100 000 years after the creation.

A more interesting question remains, however. What is the origin of the nuclei of hydrogen and helium? Are they also relics of physical processes that occurred at still earlier epochs? During the first few minutes after the big bang, the temperature of the cosmic plasma was in excess of 10^6 K, which is hot enough for nuclear reactions to occur. Using computer models combined with nuclear data, astrophysicists can reconstruct the details of the nuclear activity which took place in the first minutes of the universe.

At the end of the first second the temperature was 10^{10} K – too hot for composite nuclei to have existed. Instead, a 'soup' of individual protons and neutrons in chaotic motion filled all of space, mingled with electrons, neutrinos and photons (heat radiation). The early universe expanded extremely rapidly, so that when one minute had elapsed the temperature had dropped to a 10^8 K, and after several minutes it had fallen below the level at which nuclear reactions are possible. There was thus a relatively brief period of a few minutes during which the protons and neutrons could aggregate together into composite nuclei.

The principal nuclear reaction was the fusion of protons and neutrons to form helium nuclei, which consist of two protons and two neutrons apiece. Because protons are marginally lighter than neutrons, they existed in somewhat greater abundance, so that when the production of helium was complete, some protons remained free. Calculations show that very little else happened in the short time available; thus the composition of the emerging plasma was about 10 per cent helium nuclei and 90 per cent hydrogen nuclei, which reflects with satisfying precision the observed abundances of these elements in the universe today. The conclusion is that the element helium is a fossil remnant from the primeval furnace which raged during the first few minutes following the creation.

It is fortunate that the primeval material was over-rich in protons, for it is from the residue of unmatched protons that the universe derives its hydrogen. Without hydrogen the sun would not burn, nor would there be any water in the cosmos. It seems unlikely that life could exist in these circumstances.

Fossils from the first second

The fact that astrophysicists have identified a fossil from the first few minutes of existence is a dazzling achievement. But scientists are never satisfied, always pushing on beyond the known limits of their subject to the next goal. That is the essence of research. An explanation of the chemical elements requires a knowledge of the universe at the end of the first second. But what about earlier moments, epochs *within* the first second?

To embark on such a study is to enter an Alice-in-Wonderland world of mysterious states of matter and unfamiliar forces. It is to probe closer still towards the greatest mystery of all, the creation event itself. To help build up a picture of the universe before it was 1 s old, let us imagine that we can climb aboard a time machine that will take us back moment by moment from 1 s towards the time 0 at which the universe exploded into existence. But caution! Most of our present understanding of the first second is based on conjecture and extrapolation. Confirmatory evidence is very hard to obtain. What follows in this and subsequent chapters is a result of theoretical modelling, some of it contentious and speculative.

To make sense of the events that unfolded in the very early universe, it is necessary to understand the nature of cosmic activity. If we could journey back in time from the present day we should find that the farther back we went, the more the pace of events would quicken. Throughout the Earth's 4 600 million year history change has been slow. Geological time-scales are reckoned in millions of years. If we could journey back to a few million rather than a few thousand million years from the big bang, we should find things happening much faster. Galaxies were being formed in the space of a few hundred million years, while stars formed even more quickly, perhaps in a few tens of millions.

Back before 100 000 years the universe was relatively featureless. This is the phase of glowing plasma. The pace of events can be gauged here by the rate of cosmic expansion and the rate at which the temperature was falling. The expansion was some 100 000 times faster then than it is today. The temperature was several thousand degrees. At earlier times the rate was faster still and the temperature higher. At time 1 s the universe was doubling in size in about a second, its temperature 10^{10} K. Within the first second the pace of change escalated still faster, rising without limit as the instant of creation was approached.

Mathematically, this accelerating rate of activity is described as a 'reciprocal' relationship. For example, the expansion rate is proportional to $1/t$ and the temperature to $1/\sqrt{t}$, where t is the time since the creation. As t becomes smaller and smaller, so these quantities rise faster and faster towards infinite values. Because the level of activity rises steeply as we journey back in time to the first moment, important changes are likely to occur in briefer and briefer time intervals. It is then more meaningful to adopt a 'power of ten' approach to time. For example, as much happened during the interval from 0.1 s up to 1 s as occurred in the interval from 0.01 s up to 0.1 s. And so on. Each time the interval is further subdivided by ten so we encounter a comparable degree of change compressed into that smaller interval.

The obvious question that arises at this stage is how far back we can extrapolate our model of the early universe with any sort of confidence. I remember when I was a student in the late 1960s attending a lecture on cosmology at which the recently discovered cosmic background heat radiation was mentioned. The lecturer seemed a little embarrassed to discuss the calculations of helium abundance based on the nuclear reactions which occurred in the first few minutes. Most of the audience laughed openly at the audacity of the project, and clearly felt that modelling the universe at such an early epoch was unacceptably speculative. Today the mood has changed dramatically. The helium calculation has become part of established cosmological doctrine, and attention focusses on much earlier epochs than the nucleosynthesis phase.

It often comes as a surprise to learn that the extreme conditions which prevailed during most of the first second of the universe are within current experimental experience. Modern particle accelerators can, for a brief instant, simulate the physical conditions that occurred as early as 10^{-12} s, when the temperature was a staggering 10^{16} K and today's entire observable universe was compressed into a region no larger than the solar system. So in our journey back into the strange world of the primeval cosmos, experiment as well as theory can be our guide for part of the way.

As we probe farther and farther back, then, we encounter more and more extreme physical conditions. The most important parameter to gauge our progress is energy. The energy possessed by a typical particle as it cavorts about in the primeval 'soup' or plasma rises ever more sharply as the first moment is approached. At 1 min we have to deal with nuclear energies. At 1 s we reach energies that can be attained by some radioactive emissions. At 1 microsecond (1 μs), the energy of a typical particle is comparable to that in the early particle accelerators. When we reach 10^{-12} s (1 picosecond [1 ps]), we are approaching the frontiers of current high-energy particle physics. Beyond this point, theory is our only aid.

In earlier chapters we have seen how the four forces of nature are now considered to be parts of a single master force, the superforce. Our mistake in attributing separate identities to the four forces came about because we normally observe the world at comparatively low energies. As the energy is raised, so the forces begin to merge. First the electromagnetic force merges with the weak force. This occurs at an energy equivalent to about ninety proton masses, corresponding to a temperature of around 10^{15} K. Existing accelerators can just reach this regime, which is where W and Z particles are liberated. The further merging of the electroweak and strong forces, and eventually gravity, does not occur until enormously greater energies have been reached. We have to attain the unification and Planck scales, trillions of times more energetic than the electroweak scale.

We can now perceive the primeval universe as a huge natural laboratory in which the energy of the big bang is released to drive physical processes that are beyond all laboratory experience. Though we may never be able to experiment with the superforce directly we can appeal to cosmology to provide us with clues of its fleeting activity in the first moments of cosmic existence.

At 10^{-12} s after the big bang the temperature was so high that all the now familiar particles and antiparticles would have been created out of the available heat energy. The universe at this moment contained almost equal proportions of matter and antimatter. Later, when the particle – antiparticle pairs which constituted the majority of material annihilated, a residue of matter was left. The density of particles was so high that an equilibrium was reached, in which the available energy was partitioned democratically among all the different particle species.

The nature of the cosmic material at this stage was quite unlike anything of which we have direct experience. Being packed together so densely, the hadrons did not possess individual identities. Protons and neutrons did not exist as separate entities. Instead, the cosmic material consisted of a fluid of quarks, moving about more or less independently. Furthermore, at these energies all distinction between the weak and the electromagnetic force was lost, and the nature of the leptons and quarks was peculiar indeed. Particles such as electrons, muons, and neutrinos that we see today did not exist in their familiar forms. Photons, Ws, and Zs were inextricably conflated in identity. If we could journey back in time to this moment we would see an entirely novel phase of matter, unknown to man, in which particles had not yet arranged themselves into those that could be identified by a particle physicist.

The key to understanding this weird high-temperature phase of matter is symmetry-breaking. In Chapter 8 it was explained how a gauge symmetry can be spontaneously broken to supply particles with masses and provide the distinction between the electromagnetic and weak forces. There is a general rule in nature that high temperatures tend to restore symmetry. A good example of this rule is provided by the two phases of water, liquid and ice. A crystal of ice displays certain preferred directions in space, along the edges of the crystal lattice. When the ice melts the crystal structure breaks up. The droplet of water that replaces the ice does not display any preferred direction in space, i.e. it is symmetric. The effect of raising the temperature is to restore the underlying orientation symmetry that was spontaneously broken by the ice crystal. When matter is heated to 10^{16} K, an analogous phase change occurs, akin to that from ice to water. This time, however, the symmetry that is restored is the underlying gauge symmetry of the electroweak force.

Our picture of the universe at 1 ps (10^{-12} s) is thus a remarkable one. The universe is filled with a mysterious fluid medium, unknown anywhere in the cosmos since. The inhabitants of the universe are not particles we recognize. But this peculiar phase of matter cannot persist. As the temperature drops, a sudden phase change occurs, reminiscent of water freezing to ice. Abruptly, all the familiar particles – electrons, neutrinos, photons, quarks – are identifiable. The gauge symmetry has been broken, and the electromagnetic force separates out from the weak force.

If we follow the further progress of the cosmic material forward in time, another crucial phase change occurs at about 1 millisecond (1 ms). The tight press of agitated quarks suddenly congeals into a sea of well-defined hadrons. Individual protons, neutrons, mesons, and other strongly interacting particles can now be discerned, with the quarks linked together in distinct units of two and three. Later still, as the temperature sinks yet lower, all the remaining antiparticles, such as the positrons annihilate, producing large amounts of gamma radiation, and the cosmic material now contains the more familiar mix of protons, neutrons, electrons, neutrinos, and photons that set the stage for helium synthesis after a few seconds elapse.

Our study of the universe from 10^{-12} s onwards has produced a fascinating new perspective of the nature of matter. We can now see that protons and neutrons – the building blocks of the universe – did not always exist, but congealed from a broth of quarks at about 10^{-3} s. These nuclear particles can therefore be regarded as fossils of the first millisecond. Still more bizarre is the fact that the leptons and quarks which go to make up all matter, only achieved their present identities at about 10^{-12} s. They are fossils from the first picosecond.

A systematic picture is beginning to emerge. We can trace the origin of the elements back to distant epochs of star burning, and nucleosynthesis in the first minutes of the universe. The protons and neutrons that go to build these elements find their origin at still earlier moments, while the leptons, and the quarks that in turn build the nuclear particles, are relics of a time when the universe had existed for a mere one-million-millionth of a second. But a mystery still remains, a mystery that takes us back to an epoch much earlier still – to the so-called GUT era.

The origin of matter

When the big bang theory was first proposed, no convincing explanation was forthcoming for how the material which erupted from the primeval explosion first came to exist. Cosmologists had to fall back on the assumption that the matter from which the universe is constructed was present in the beginning. No physical process known at the time could bring this matter into existence. Today, the new cosmology provides a very plausible explanation for the origin of matter, based on the activities of the superforce.

The possibility that matter may be created out of concentrated energy has been known for several decades. And there was no lack of energy in the big bang to generate all the matter in the visible universe – about 10^{50} tonnes in total. The mystery is how all this matter came to exist without an equal quantity of antimatter, a problem mentioned briefly in Chapter 2. When matter is created in the laboratory, antimatter is always produced too, and the symmetry between matter and antimatter seems to be deep-rooted in the laws of physics. This raises the obvious question of where all the antimatter has gone.

First of all we have to be sure that the universe really is made entirely of matter. A rock made of antimatter would look in all respects like a rock made of matter. You could not tell them apart by looking. There is, however, a most unmistakable way to determine which is which. If each of them is brought into contact with a piece of matter, the antimatter rock will vanish amid an explosive outburst of nuclear proportions. Even a tenuous spray of gas would cause the antimatter to react violently, spewing forth intense gamma radiation. Clearly, we can be sure that the Earth is made of 100 per cent matter.

But is this asymmetry true of the universe as a whole? As far as we can tell it is. If our galaxy contained substantial quantities of antimatter, the inevitable collisions that occur between gas, dust, stars, planets, and other objects would produce a deluge of gamma radiation as the antimatter encountered matter and annihilated itself. Gamma radiation at this intense level would certainly be detected, and astronomers have placed a limit on the antimatter content of our galaxy of one part in 10^3. Apart from the occasional antiprotons found in cosmic rays, the galaxy seems to be pretty well pure matter.

It is conceivable that some other galaxies are made almost entirely of antimatter, with very little matter. However, even galaxies collide from time to time, and in the past they were much closer together. The gamma rays from these encounters would be detectable today. Moreover, taking the universe as a whole, it is hard to see how an initial mixture of matter and antimatter could ever separate into disconnected regions of space. Given this weight of evidence, most cosmologists believe the cosmos is made predominantly of matter, and that this asymmetry has been locked into the universe since its earliest moments.

Ten years ago the only explanation offered for the primordial imbalance between matter and antimatter was to suppose that it was built in from the beginning, that the material issuing from the big bang had a disproportionate amount of matter as compared to antimatter. This type of 'explanation' – the appeal to contrived initial conditions – falls into the category of saying that things are what they are because they were what they were. It cannot be called science. Virtually any composition of the primordial material could be explained in the same way. It gives us no idea why the imbalance was as little or as great as it was. There seems to be no good reason why, say, twice or even a million times as much matter was not created.

GUTs to the rescue

Rather than assuming that the excess of matter in the universe is god-given, a more satisfying explanation is to suppose that initially there was complete symmetry between matter and antimatter and that, somehow, a preponderance of matter developed *after* the beginning by natural causes, and then became 'frozen in' to the universe. It would then no longer be necessary to believe in an arbitrary initial condition; the state of exact equality (zero excess) is unique. The observed excess of matter over antimatter might then be explained quantitatively in terms of a physical theory.

For this idea to work it is obviously necessary to have a physical mechanism that breaks the matter–antimatter symmetry, traditionally one of the inviolable rules of physics. In the late 1970s just such a symmetry-breaking mechanism came to hand in the guise of the grand unified theories (GUTs). As explained in earlier chapters, one of the more sensational predictions of GUTs is that protons are unstable, and decay into positrons. The relationship between proton decay and matter–antimatter asymmetry can be seen by considering the possible long-term fate of a hydrogen atom (a proton plus an electron). When the proton decays, it emits a pion and a positron. The pion decays into two photons, while the positron can annihilate the electron to give two more photons. What started out as an atom of matter ends up as pure radiation energy. By this process, matter has been completely converted into energy, without encountering antimatter. Now every physical process can be reversed, which means that it is possible to conceive of energy turning into matter without the production of antimatter. This process, greatly speeded up, could explain how matter came to exist.

To model the creation process in detail, it is necessary to go back to the so-called GUT era, a full twenty powers of ten beyond the electroweak epoch that we considered in the previous section. This means attempting to describe a universe only 10^{-32} s old! At that instant the cosmos would be filled with a broth of weird, unrecognizable particles, some of them extraordinarily heavy, packed to a density of 10^{73} kg m^{-3}, and bathed in heat at a temperature of 10^{28} K. The universe was at that time so young that light could not have made its way the equivalent of one-thousand-millionth of the distance across a proton since the first moment.

The crucial ingredients of this exotic broth are the superheavy particles that transmit the grand unified force, the so-called X particles mentioned in Chapter 8. It is these particles which can introduce the lopsidedness between matter and antimatter. This is how: when an X decays it may yield many daughter particles of which, say, ⅔ may be matter but only ⅓ antimatter. The precise details of this asymmetry depend on the particular GUT adopted, but clearly X decay can lead to a preponderance of matter.

There is one point to take care of though. The primordial broth will also contain the antiparticles of X, usually denoted $\bar{X}$. Our assumption, remember, is that the universe started out symmetrically, so there would have been equal proportions of X and $\bar{X}$. When the $\bar{X}$s decay, they reverse the asymmetry, yielding ⅔ antimatter to ⅓ matter. The net effect is that the initial symmetry remains intact.

To escape from this impasse, theorists assume that there must be a fundamental imbalance in the decay rates of X and $\bar{X}$. As a result, the $\bar{X}$ decays do not quite offset the X decays. There is perhaps a bias of about one in a thousand million in favour of X, giving us a one-in-a-thousand-million preponderance of matter over antimatter.

How reasonable is this assumption? Physicists are keen historians, especially when it comes to their own subject. The lessons of history are never far from their minds when it comes to constructing new theories. One such lesson occurred in 1956. Two Chinese-American physicists, T. D. Lee and C. N. Yang, upturned the apple-cart by insisting that the weak force violates a formerly sacrosanct symmetry of nature, known as mirror symmetry. Until that time, physicists had assumed, more or less without thinking, that the forces of nature were indifferent to the distinction between left and right. To be sure, there are many examples of natural structures with an inbuilt 'handedness', the most famous being DNA. Molecules of DNA have a shape like a spiral staircase winding always to the right. Though no left-handed DNA occurs naturally, there is no fundamental law of physics that prevents this. The fact that all Earth life is made from the right-handed form is presumably because the first self-replicating molecule just happened to be twisted that way. It is a good example of spontaneous symmetry-breaking; the actual structure is asymmetric even though the underlying physical forces remain symmetric.

When a physicist says that the forces of nature are mirror-symmetric he means that the fundamental processes induced by the forces would look equally permissible in a mirror as they do when viewed directly. Imagine taking a movie film of the decay of some particle, and loading the film back to front in the projector. If the forces that drive the decay are mirror-symmetric, no physicist could spot the deception.

So confident were physicists that subatomic particles could not tell left from right that they did not think it was necessary to test the matter. Then along came Lee and Yang who challenged the assumption. An experiment was quickly performed by another Chinese-American, Mrs C. S. Wu, and to everybody's consternation it was discovered that Lee and

Yang were right. The weak force does indeed violate mirror-symmetry. The Wu experiment, which involved measuring the numbers of left- and right-moving electrons emitted by carefully aligned radioactive cobalt nuclei, proved to be a turning point in physics. After that, no symmetries were safe.

In 1964 there followed a second shock. A lot of interest was being taken in the intriguing behaviour of a peculiar particle called the neutral K meson. The idea that mirror symmetry was violated had by then long been accepted, but it was assumed that antiparticles would always violate mirror symmetry in the reverse sense to particles. (Antiparticles usually display the reverse properties of particles.) If this were in fact the case, there would be no way that the universe could generate a preponderance of matter over antimatter in the big bang, because for any process in which a particle was created there would be another mirror process somewhere in which an antiparticle was created. The peculiarities of the neutral K meson, which is a sort of hybrid particle–antiparticle, made it possible to test these ideas.

The crucial experiment was performed by V. L. Fitch and J. W. Cronin at the Brookhaven National Laboratory. They found that mirror symmetry is *not* violated in equal and opposite ways by particles and antiparticles, at least as far as the K meson is concerned. There is a tiny, but highly significant lopsidedness here too. Such an asymmetry reflects a fundamental imbalance in the nature of the forces that drive some particle decays, providing concrete experimental evidence for asymmetry between matter and antimatter.

In the late 1970s theorists began modelling the GUT phase of the big bang on the assumption that the above asymmetry really does exist in the grand unified force, and came up with numbers that suggest, typically, an imbalance between matter and antimatter of one in a thousand million. This means that for every thousand million antiparticles, a thousand million plus one particles are created. Although only tiny, the slight excess of particles proves to be absolutely crucial. When the universe eventually cools, the antimatter annihilates, and in so doing it destroys nearly all the matter. But not quite all, as there is that one part in a thousand million excess of matter over antimatter that remains left over. It is from this minute residue – almost an afterthought of nature – that all the objects in the universe, including us, are made. So ultimately, *all* matter is a fossil, a relic from the GUT era, a mere 10^{-32} s from the creation event.

If this analysis is to be believed, the overwhelming majority of matter that emerged from the big bang disappeared before the first few seconds had elapsed, along with all the cosmic antimatter. Now we know why there is so little antimatter in the universe. But this vanished material has left an echo of its erstwhile existence in the form of energy. The matter–antimatter annihilation produced about a thousand million gamma ray photons for each electron and each proton that remained unscathed. Today this radiation has been cooled by the cosmic expansion, and forms the background heat radiation that fills the universe. Apart from the energy locked up in matter, this background heat accounts for the greater part of the energy of the universe. We thus have at hand a theory that not only explains how matter came to exist, but can also account for the ratio of matter to energy in the universe.

Before the advent of GUTs, the temperature of the cosmic background heat radiation could not be explained. The level of radiant heat energy was just another apparently arbitrary cosmic parameter, built into the universe at its creation. No reason was known why the temperature today could not be 0.3 or 30 K, rather than the 3 K it is. The grand unified theories provide a means to explain that temperature from physics. A present temperature of 3 K corresponds to about 10^9 photons to every proton and electron in the universe, and this value is in good agreement with the typical one-in-a-thousand-million excess of particles over antiparticles predicted by GUTs. One of the fundamental parameters of cosmology can thus be explained in terms of physical processes which occurred during the GUT era. It was at that unimaginably early moment of existence that the foundations were laid for the structure of the universe we see today.

12
What Caused the Big Bang?

The genesis paradox

Whenever I give a lecture on cosmology one question never fails to be asked: What caused the big bang? A few years ago I had no real answer. Today, I believe we know what caused the big bang.

The question is actually two rolled into one. We should like to know why the universe began with a bang, what triggered this explosive outburst in the first place. But behind this physical enigma lies a deeper metaphysical mystery. If the big bang represents the origin of physical existence, including that of space and time, in what sense can anything be said to have *caused* this event?

On a purely physical level, the abrupt appearance of the universe in a huge explosion is something of a paradox. Of the four forces of nature which control the world, only gravity acts systematically on a cosmic scale, and in all our experience gravity is attractive. It is a pulling force. But the explosion which marked the creation of the universe would seem to require a pushing force of unimaginable power to blast the cosmos asunder and set it on a path of expansion which continues to this day.

People are often puzzled in the belief that if the universe is dominated by the force of gravity it ought to be contracting, not expanding. As a pulling force, gravity causes objects to implode rather than explode. For example, a highly compact star will be unable to support its own weight, and may collapse to form a neutron star or a black hole. In the very early universe, the compression of material exceeded that of even the densest star, and this fact often prompts the question of why the primeval cosmos did not itself turn into a black hole at the outset.

The traditional response leaves something of a credibility gap. It is argued that the primeval explosion must simply be accepted as an initial condition. Certainly, under the influence of gravity, the rate of cosmic expansion has continually slowed since the first moment, but at the instant of its creation the universe was expanding infinitely rapidly. No force caused it to explode in this way, it simply started with an initial expansion. Had the explosive vigour been less extreme, then gravity would very soon have overwhelmed the dispersing material, reversing the expansion and engulfing the entire cosmos in a catastrophic implosion, producing something rather like a black hole. As it happened, the bang was big enough to enable the universe either to escape its own gravity and go on expanding for ever under the impetus of the initial explosion, or at least to survive for many thousands of millions of years before succumbing to implosion and annihilation.

The trouble with this traditional picture is that it is in no sense an explanation for the big bang. Once again, a fundamental feature of the universe is merely attributed to an *ad hoc* initial condition. The big bang 'just happened'. We are left uncomprehending as to why the force of the explosion had the strength that it did. Why did the universe not explode more violently still, in which case it would be expanding much faster today? Alternatively, why is it not expanding much slower, or even contracting by now? Of course, had the cosmos *failed* to explode with sufficient violence, and rapid collapse overtaken it, we should not be here to ask such questions; but that is hardly an explanation.

Closer investigation shows that the genesis paradox is actually deeper than this. Careful measurement puts the rate of expansion very close to a critical value at which the universe will just escape its own gravity and expand for ever. A little slower, and the cosmos would collapse, a little faster and the cosmic material would have long ago completely dispersed. It is interesting to ask precisely how delicately the rate of expansion has been 'fine-tuned' to fall on this narrow dividing line between two catastrophes. If at time 1 s (by which time the pattern of expansion was already firmly established) the expansion rate had differed from its actual value by more than 10^{-18}, it would have been sufficient to throw the delicate balance out. The explosive vigour of the universe is thus matched with almost unbelievable accuracy to its gravitating power. The big bang was not, evidently, any old bang, but an explosion of exquisitely arranged magnitude. In the traditional version of the big bang theory we are asked to accept not only that the explosion just happened, but that it happened in an exceedingly contrived fashion. The initial conditions had to be very special indeed.

The rate of expansion is only one of several apparent cosmic 'miracles'. Another concerns the pattern of expansion. As we observe it today, the universe is extraordinarily uniform on the large scale, in the way that matter and energy are distributed. From the viewpoint of a distant galaxy, the overall structure of the cosmos would appear almost identical to its aspect from Earth. The galaxies are scattered throughout space with a constant average density, and at every point the universe would look the same at all orientations. The primeval heat radiation which bathes the universe arrives at Earth with a uniform temperature in every direction accurate to one part in ten thousand. This radiation has travelled to us across thousands of millions of light years of space, and would carry the imprint of any departures from uniformity encountered on the way.

The large-scale uniformity of the universe continues to be preserved with time as the universe expands. It follows that the expansion itself must be uniform to a very high degree. Not only is the rate of expansion the same in all directions, it is the same from region to region within the cosmos. If the universe were to expand faster in one direction than the others, it would depress the temperature of the background heat radiation coming from that direction, and also distort the pattern of motion of the galaxies as viewed from Earth. So not only did the universe commence with a bang of a quite precise magnitude, it was a highly orchestrated explosion as well, a simultaneous outburst of exactly uniform vigour everywhere and in every direction.

The extreme improbability that such a coherent, synchronized eruption would occur spontaneously is exacerbated by the fact that, in the traditional big bang theory, the different regions of the primeval cosmos would have been causally isolated. The point here is that, on account of the theory of relativity, no physical influence can propagate faster than light. Consequently, different regions of the universe can come into causal contact only after a period of time has elapsed. For example, at 1 s after the initial explosion, light can have travelled at most one light-second which is 300 000 km. Regions of the universe separated by greater than this distance could not, at 1 s, have exercised any influence on each other. But at that time, the universe we observe today occupied a region of space at least 10^{14} km across. It must therefore have been made up of some 10^{27} causally separate regions, all of them nevertheless expanding at exactly the same rate. Even today, when we observe the cosmic heat radiation coming from opposite sides of the sky, we are receiving identical thumbprints from regions of the universe that are separated from each other by ninety times the distance that light could have travelled at the time the heat radiation was emitted towards us.

How is it possible to explain this remarkable degree of co-operation between different parts of the universe that apparently have never been in communication with each other? How have they come to behave so similarly? The traditional response is, yet again, to fall back on special initial conditions. The extreme uniformity of the primeval explosion is simply regarded as a brute fact: 'The universe began that way.'

The large-scale uniformity of the universe is all the more mysterious on account of the fact that, on a somewhat smaller scale, the universe is *not* uniform. The existence of galaxies and galactic clusters indicates a departure from exact uniformity, a departure which is, moreover, of the same magnitude and scale everywhere. Because gravity tends to amplify any initial clumping of material, the degree of non-uniformity required to produce galaxies was far less during the big bang than it is today. In spite of this, some small degree of irregularity must have been present in the primeval phase or galaxies would never have started to form. In the old big bang theory these early irregularities were also explained away as initial conditions. Thus, we were required to believe that the universe began in a peculiar state of extraordinary but not quite perfect order.

The explanation can be summarized as follows: with gravitational attraction the only cosmic force available, the big bang must simply be accepted as god-given, an event without a cause, an assumed initial condition. Furthermore, it was an event of quite astonishing fidelity, for the present highly structured cosmos could not have arisen unless the universe was set up in just the right way at the outset. This is the genesis paradox.

The search for antigravity

Although a resolution of the genesis paradox has been achieved only in the past few years, traces of the essential idea go back far into history, to a time before the expansion of the universe or the big bang theory were known. Even Newton realized that there was a deep puzzle about the stability of the cosmos. How can the stars just hang out there in space unsupported? The universal force of gravity, being attractive, ought to cause the entire collection of stars to plunge in on itself.

To escape from this absurdity, Newton used a curious argument. If the universe collapsed under its own gravity, he reasoned, every star would be obliged to fall towards the centre of the stellar assemblage. But suppose the universe were infinite, with stars distributed on average uniformly throughout infinite space. There would then be no overall centre towards which the stars could fall; in an infinite universe every region is identical to every other. Any given star would receive gravitational pulls from all its neighbours, but these would average out in their different directions, and so there would be no systematic net force to convey a star towards any particular place of general congregation.

When Einstein replaced Newton's gravitation theory 200 years later, he was equally troubled by the enigma of how the universe avoids collapse. His first paper on cosmology was published before Hubble's famous discovery of the expanding universe and Einstein presumed, like Newton, that the cosmos was static. His solution of the stability problem was, however, much more direct. Einstein believed that to prevent the universe from imploding under its own gravity there had to be another cosmic force to counteract the gravitational force. This new force would have to be repulsive rather than attractive, a pushing force to balance the pull of gravity. In this respect it might be regarded as 'antigravity', although 'cosmic repulsion force' is a more accurate description. Einstein did not conjure up the cosmic repulsion force in an *ad hoc* way. He found that his gravitational field equations contained an optional term which gave rise to a force with exactly the desired properties.

Although the idea of a repulsive force pushing against the gravity of the universe is simple enough to grasp in broad conception, the actual properties of the force are decidedly odd. It goes without saying that we do not notice any such force on the Earth, nor has any hint of one been found during several centuries of planetary astronomy. Evidently, if a cosmic repulsion force exists, it must have the property that it does not act conspicuously at close range but accumulates in strength over astronomical distances. Behaviour of this sort was contrary to all previous experience of forces, which tend to be strong nearby and to weaken with distance. Electric and gravitational forces, for example, fall steadily towards zero in accordance with the inverse square law. Nevertheless, a force of this somewhat peculiar sort emerged naturally from Einstein's theory.

The cosmic repulsion found by Einstein should not really be thought of as a fifth force of nature. Rather, it is a weird offshoot of gravity itself. In fact the effects of cosmic repulsion can be attributed to ordinary gravity if the source of the gravitational field is chosen to be a medium with rather unusual properties. A familiar material medium, such as a gas, will exert a pressure, but the hypothetical cosmic medium being discussed here is required to possess a *negative* pressure, or tension. To get some idea of what is involved, imagine that we could fill a vessel with this conjectured cosmic stuff. Instead of pushing against the walls of the vessel like an ordinary gas, the cosmic medium would try to pull the walls inwards.

We can envisage the repulsion, therefore, either as a sort of adjunct of gravity, or as caused by the ordinary gravity of an invisible fluid medium with negative pressure filling all of space. There is no conflict, incidentally, between the fact that negative pressure would suck on the walls of a vessel, and the fact that the hypothetical medium exerts a repulsion on galaxies rather than an attraction. The repulsive effect is due to the *gravity* of the medium, not its mechanical action. In any case, mechanical forces arise from pressure differences, not pressure as such and the medium is supposed to fill all of space. It could not be confined to a vessel. In fact, an observer immersed in the medium would not perceive any tangible substance at all. Space would look and feel completely empty.

In spite of these rather bizarre features, Einstein duly declared that he had a convincing model of a universe held in equilibrium between the attractive force of gravity and the newly discovered cosmic repulsive force. With the aid of simple arithmetic he calculated the strength needed for the repulsive force to balance the gravity of the universe. Einstein was able to confirm that the repulsion would be so slight within the solar system, and even the galaxy, that we would never have spotted it observationally. For a while it seemed that an age-old puzzle had been brilliantly solved.

Then things began to go wrong. First there was a problem about stability. The essential idea was to match the forces of attraction and repulsion precisely. But like many balancing acts this one turns out to be a delicate affair. If, for example, Einstein's static universe were to expand a fraction, the attractive force of gravity (which diminishes with separation) would go down a bit, while the cosmic repulsion force (which increases with distance) would go up. This would lead to an overbalance, with the repulsion winning out and forcing a still greater expansion, and leading to the eventual runaway distension of the universe under an all-dominating repulsion. On the other hand, if the universe shrank a little, the gravitational force would go up and the repulsion would go down, causing gravity to win out, and the universe would then shrink faster and faster towards the total collapse that Einstein had sought to avoid. Thus, the slightest hiccup, and the carefully balanced equilibrium would fail, spelling cosmic disaster.

Then, in 1927, Hubble discovered the expansion of the universe. All balancing acts were thereby rendered obsolete. It was immediately apparent that the universe avoids *implosion* because it is engaged in *explosion*. Had Einstein not been sidetracked with the repulsive force he would surely have made this deduction theoretically and thus have predicted the expansion of the universe a decade before its discovery by astronomers. That would surely have gone down in history as one of the great theoretical predictions of all time. In the event, Einstein abandoned the cosmic repulsion force in disgust. 'The biggest blunder of my life,' he was later to bemoan. But that was by no means the end of the story.

Cosmic repulsion was invented by Einstein to solve a non-existent problem, namely how to explain a static universe. But as with all genies, once this one was let out of the bottle nobody could put it back again, and the possibility that the dynamics of the universe is a competition between attractive and repulsive forces lived on. Although astronomical observations do not reveal cosmic repulsion at work, they cannot prove it is non-existent. It may simply be too weak to have shown up yet.

Einstein's field equations, though admitting a repulsive force in a natural way, make no restriction on the *strength* of the force. Einstein was free to postulate, after his bitter experience, that the strength was precisely zero, thus eliminating the repulsion altogether. But there was no compelling reason to do this. Other scientists were happy to retain the repulsion, even though it was no longer needed for its original purpose. In the absence of evidence to the contrary, they reasoned, nobody is justified in setting the force to zero.

The consequences of retaining the repulsion force in the expanding universe scenario are easily worked out. Early in the life of the cosmos, when the universe is compressed, the repulsion can be ignored. During this phase, the effect of the gravitational attraction is to slow the pace of expansion, in the same way that a missile fired vertically is slowed by the Earth's gravity. If it is assumed without explanation that the universe starts out expanding rapidly, then gravity acts to steadily reduce the rate to the value observed today. With time, the gravitational force weakens as the cosmic material disperses. By contrast, the cosmic repulsion grows because the galaxies move farther apart. Eventually, the repulsion force comes to exceed the gravitational attraction, and the expansion rate begins to pick up again, getting faster and faster. Thereafter, the universe is dominated by the cosmic repulsion and spends all of eternity in runaway expansion.

Astronomers have reasoned that this unusual behaviour, in which the universe first slows up and then accelerates again, ought to be evident in the observed motion of the galaxies. Careful astronomical observations have failed to provide any convincing evidence for such a turn-about, though from time to time claims have been made to the contrary.

Curiously, the idea of a universe caught by runaway expansion had already been mooted by the Dutch astronomer Wilhelm de Sitter in 1916, several years before Hubble discovered that the universe was expanding. de Sitter argued that if the universe were devoid of ordinary matter, then the usual attractive force of gravity would be absent, and the cosmos would come under the sole influence of the repusion. This would make the universe expand. (At the time this was a novel idea.)

To an observer, who would be unable to see the curious invisible fluid medium with negative pressure, it would merely appear as if empty space were expanding. The expansion could be judged by stationing test bodies in various positions and watching them move apart. The idea of expanding empty space was considered little more than a curiosity at the time, although it turned out to be remarkably prophetic, as we shall see.

What can be concluded from this saga? The fact that astronomers do not see a cosmic repulsive force at work does not logically imply that the force is non-existent. It could be that it is too weak to be detected with present instrumentation. All observations contain a level of error, and only an upper limit on the strength of the force can be obtained. Set against this, it could be argued on aesthetic grounds that the laws of nature would be simpler if cosmic repulsion were absent. This inconclusive debate about the existence of 'antigravity' had been grinding on for many years when suddenly an entirely new twist occurred that gave the subject an unexpected immediacy.

Inflation: the big bang explained

In the previous section we saw how, if a cosmic repulsive force exists, it must be very weak and far too weak to have had any significant effect on the big bang. But this conclusion rests on the assumption that the strength of the repulsive force does not change with time. In Einstein's day everybody made this assumption because the force was put into the theory 'by hand'. No one considered the possibility that cosmic repulsion might be *generated* by other physical processes that could change as the universe expands. Had such a possibility been entertained, then the history of cosmology would have been very different, for one could then conceive of a scenario in which, under the extreme conditions of the early universe, cosmic repulsion momentarily dominated gravity causing the universe to explode, before fading into insignificance.

This general scenario is precisely what has come out of recent work on the behaviour of matter and forces in the very early universe. It is now clear that a huge cosmic repulsion is an inevitable by-product of the activities of the superforce. The 'antigravity' that Einstein threw out of the door has come back in through the window.

The key to understanding the re-discovered cosmic repulsion is the nature of the quantum vacuum. We have seen how such a repulsion can be produced by a bizarre invisible medium which looks identical to empty space but which possesses a negative pressure. Physicists now believe that that is exactly how a quantum vacuum would be.

In Chapter 7 it was emphasized how the vacuum must be regarded as a ferment of quantum activity, teeming with virtual particles and full of complex interactions. It is important to appreciate that, at the quantum level of description, the vacuum is the dominant structure. What we call particles are only minor disturbances bubbling up over this background sea of activity.

In the late 1970s it became apparent that the unification of the four forces required a drastic re-appraisal of the physical nature of the vacuum. The theory suggested that all this vacuum energy could arrange itself in more than one way. To put it simply, the vacuum could become excited and adopt a number of states of very different energy, in the same way that an atom can be excited to higher energy levels. These several vacuum states would look identical if we could view them, but they possess very different properties.

First of all, the energy involved leaps by huge amounts from one vacuum state to another. In the grand unified theories, to take an example, the gap between the least and greatest vacuum energy is almost incomprehensibly large. To get some feeling for the enormity of the numbers involved, consider the huge outpouring of energy from the sun, accumulated over its entire lifetime of about 5 thousand million years. Conceive of taking this colossal quantity of energy – the entire output of the sun during its whole history – and compressing it into a volume of space less than that occupied by the solar system. You then begin to approach the sort of energy density contained in a GUT vacuum state.

Alongside these staggering energy differences are equally enormous changes in the pressure of the vacuum states. But here comes the important twist: the pressures are all *negative*. The quantum vacuum behaves exactly like the previously hypothetical medium which produces cosmic repulsion, only this time the numbers are so big that the strength of the repulsive force is 10^{120} times greater than Einstein needed to prop up a static universe.

The way now lies open for an explanation of the big bang. Suppose that, in the beginning the universe found itself in an excited vacuum state (physicists call this a 'false' vacuum). In this state the universe would be subject to a cosmic repulsion force of such magnitude that it would cause headlong expansion at a huge rate. In fact, during this phase, the universe would resemble de Sitter's model mentioned in the previous section. The difference is that, whereas de Sitter envisaged a universe sedately expanding over an astronomical time-scale, the de Sitter phase driven by the false quantum vacuum is far from sedate. A typical region of space would double in size every 10^{-34} s or so!

The way in which this hyper-expansion proceeds is distinctive: distances increase in size exponentially fast. (We met the concept of exponential change in Chapter 4.) This means that every 10^{-34} s every region of the universe doubles its size, and then goes on doubling again and again in a progression. This type of runaway expansion has been dubbed 'inflation' by Alan Guth of MIT, who invented the idea in 1980. Under the impact of the exceedingly rapid and accelerating expansion, the universe would have soon have found itself swelling explosively fast. This was the big bang.

Somehow, the inflationary phase has to terminate. As with all excited quantum systems, the false vacuum is unstable and will tend to decay. When that happens, the repulsion force disappears. This would have put a stop to inflation, bringing the universe under the control of ordinary, attractive gravity. The universe would have continued to expand, of course, from the initial impetus imparted by the inflationary episode, but at a steadily falling rate. The only trace that now remains of the cosmic repulsion is this dwindling expansion.

According to the inflationary scenario, the universe started out in a vacuum state, devoid of matter or radiation. Even if matter and radiation were present initially, all traces would soon have been eradicated because the universe swelled by such an enormous factor during the inflationary phase. During this incredibly brief phase, the region of space which today forms the entire observable universe grew from one-thousand-millionth of the size of a proton to several centimetres. The density of any pre-existing material would have fallen essentially to zero.

At the end of inflation, then, the universe was empty and cold. As soon as inflation ceased, however, the universe was suddenly filled with intense heat. This flash of heat which illuminated the cosmos owed its origin to the huge reserves of energy locked up in the false vacuum. When the false vacuum decayed, its energy was dumped in the form of radiation, which instantly heated the universe to about 10^{27} K, hot enough for GUT processes to occur. From this point on the universe evolved according to the standard hot big bang theory. The heat energy created matter and antimatter, the universe began to cool, and in a succession of steps all the structure we observe today began to 'freeze' out.

The thorny problem of what caused the big bang is therefore solved by the inflationary theory: empty space itself exploded under the repulsive power of the quantum vacuum. But an enigma still remains. The colossal energy of the primeval explosion – the energy that went to generate all the matter and radiation we now see in the universe – surely had to come from somewhere? We will not have explained the existence of the universe until we have traced the source of the primeval energy.

The cosmic bootstrap

The universe came into existence amid a huge burst of energy. This energy survives in the background heat radiation and in the cosmic material – the atoms which make up the stars and planets – as 'mass' or locked-up energy. It also lives on in the outward rush of the galaxies and in the swirling activities of all the astronomical bodies. The primeval energy wound up the nascent universe and continues to drive it to this day.

Whence came this vital energy which triggered our universe into life? According to the inflationary theory the energy came out of empty space, out of the quantum vacuum. But is this a fully satisfactory answer? We can still ask how the vacuum acquired the energy in the first place.

When we ask where the energy came from we are making an important assumption about the nature of energy. One of the fundamental laws of physics is the law of *conservation* of energy, which says that although you can change energy from one form to another, the total quantity of energy stays fixed. It is easy to think of examples where this law can be tested. Suppose you have a motor and a supply of fuel, and the motor is used to drive an electric generator which in turn powers a heater. When the fuel is expended, its stored chemical energy will have been converted, via electrical energy, into heat energy. If the motor had been used instead to haul a weight to the top of a tower, and the weight were then released, on impact with the ground it would generate the same amount of heat energy as you would have obtained using the heater. The point is that however you move it about or change its form, energy apparently cannot be created or destroyed. It is a law used by engineers every day.

If energy cannot be created or destroyed, how did the primeval energy come to exist? Was it simply injected at the beginning of time, another *ad hoc* initial condition? If so, why does the universe contain the amount of energy that it does? There are about 10^{68} joules of energy in the observable universe; why not 10^{99} or $10^{10\,000}$ or any other number?

The inflation theory is one possible scientific (as opposed to metaphysical) answer to this mystery. According to the theory, the universe started out with essentially zero energy, and succeeded in conjuring up the lot during the first 10^{-32} s. The key to this miracle lies with a most remarkable fact about cosmology: the law of conservation of energy *fails* in its usual sense when applied to the expanding universe.

In fact, we have already encountered this point. The cosmological expansion causes the temperature of the universe to fall. The radiant heat energy that was so intense in the primeval phase had dwindled to a temperature close to absolute zero. Where has all that heat energy gone? The answer is, that in a sense it has depleted itself by helping the universe to expand, adding its pressure to the explosive violence of the big bang. When an ordinary fluid expands, its pressure pushes outwards and does work, so using up its energy. This means that if you expand an ordinary gas, its internal energy must fall to pay for the work done. In stark contrast to this conventional behaviour, the cosmic repulsion behaves like a fluid with *negative* pressure. When a negative-pressure fluid is expanded, its energy goes *up* rather than down. This is precisely what happened in the inflationary period, when the cosmic repulsion drove the universe into accelerated expansion. All the while the total energy of the vacuum kept on rising until, at the cessation of the inflationary era, it had accumulated to a huge amount. As soon as inflation stopped, this energy was released in a single great burst, generating all the heat and matter that eventually emerged from the big bang. From then on, the conventional positive-pressure expansion took over, and the energy began to decline again.

The creation of the primeval energy has an air of magic to it. The vacuum, with its weird negative pressure, seems to have a truly incredible capability: on the one hand it produces a powerful repulsive force, bringing about its own accelerating expansion; on the other hand, that very expansion goes on boosting the energy of the vacuum more and more. The vacuum essentially pays itself vast quantities of energy. It has an inbuilt instability to continue expanding and generating unlimited quantities of energy for free. Only the quantum decay of the false vacuum puts a stop to the bonanza.

The vacuum is nature's miraculous jar of energy. There is in principle no limit to how much energy can be self-generated by inflationary expansion. It is a revolutionary result at total variance with the centuries-old tradition that 'nothing can come out of nothing', a belief that dates at least from the time of Paremenides in the fifth century B.C. The idea of creation from nothing has, until recently, belonged solely to the province of religion. Christians have long believed that God created the universe out of nothing, but the possibility that all the cosmic matter and energy might appear spontaneously as a result of purely physical processes would have been regarded as utterly untenable by scientists only a decade ago.

For those who feel uncomfortable with the whole concept of something for nothing, there is an alternative way of looking at the creation of energy by the expanding universe. Because gravitational forces are normally attractive, it is necessary to do work to pull matter apart against its own gravity. This means that the gravitational energy of a collection of bodies is negative; if more bodies are added to the system, energy is released and the gravitational energy becomes more negative to pay for it. In the context of the inflationary universe, the appearance of heat and matter could be viewed as exactly compensated by the negative gravitational energy of the newly created mass, in which case the total energy of the universe is zero, and no net energy has appeared after all! Attractive though this way of looking at the creation may be, it should not be taken too seriously because the whole concept of energy has dubious status as far as gravity is concerned.

The antics of the vacuum are reminiscent of the story, much beloved of physicists, about the boy who falls into a bog and escapes by pulling himself up by his own bootstraps. The self-creating universe is rather like this boy since it too pulls itself up 'by its own bootstraps': entirely from within its own physical nature, the universe infuses itself with all the energy necessary to create and animate matter, driving its own explosive origin. This is the cosmic bootstrap. We owe our existence to its astonishing power.

Successes of inflation

Once the basic idea had been mooted by Guth that the universe underwent an early period of extremely rapid expansion, it became apparent that the scenario provides an elegant explanation for many of the previously *ad hoc* features of big bang cosmology.

In an earlier section we encountered several 'fine-tuning' paradoxes relating to the way that the primeval explosion was apparently highly orchestrated and precisely arranged. One of these remarkable 'coincidences' related to the way in which the strength of the explosion was exactly matched to the gravitational power of the cosmos such that the expansion rate today lies very close to the borderline between re-collapse and rapid dispersal. A crucial test of the inflationary scenario is whether it produces a big bang of this precisely matched magnitude. It turns out that because of the nature of exponential expansion – the characteristic feature of the inflationary phase – the explosive power is indeed automatically adjusted to yield exactly the right value corresponding to the universe just escaping its own gravity. Inflation can give no other expansion rate than the one that is observed.

A second major puzzle relates to the large-scale uniformity of the universe. This too is immediately explained by inflation. Any irregularities initially present in the universe would have been stretched to death by the enormous distension, rather like the wrinkles in a deflated balloon are smoothed out by inflation. With regions of space being expanded by factors of 10^{50}, any prior disorder would be diluted to insignificance.

We have seen, however, that *complete* uniformity would be incorrect, because a small degree of clumping was necessary in the early universe to account for the present existence of galaxies and galactic clusters. The original hope of astronomers was that the existence of galaxies might be explained as a result of gravitational aggregation since the big bang. A cloud of gas will tend to contract under its own gravity and then fragment into smaller clouds, which in turn fragment into still smaller clouds, and so on. It is possible to imagine the gas emerging from the big bang uniformly distributed, but purely by chance accumulations becoming overdense here and there and underdense elsewhere. Gravity would reinforce this tendency, causing the enhanced regions to grow stronger and suck in more material, and then to shrink and successively fragment, with the smallest fragments becoming stars. One would then end up with a hierarchy of structure, with stars clustered into groups, which in turn cluster into galaxies and galactic clusters.

Unfortunately, the growth of galaxies by this mechanism would take much longer than the age of the universe if there were no irregularities present in the gas at the outset, because the shrinking and fragmenting process is in competition with the expansion of the universe, which tries to disperse the gas. In the old version of the big bang theory it was necessary to assume that the seeds of galaxies were already built into the structure of the universe when it was created. Moreover, these initial irregularities had to be of just the right magnitude: too small and galaxies would never form, too large and the overdense regions would collapse into huge black holes instead. We are at a loss to know why the galaxies are the sizes they are, or why the clusters contain the numbers of galaxies that they do.

It is now possible to conceive of a better explanation for galactic structure based on the inflationary scenario. The essential idea is simple enough. Inflation occurs while the quantum state of the universe is hanging in the unstable 'false' vacuum state. Eventually the false vacuum decays, its excess energy going into heat and matter. At this point the cosmic repulsion disappears and inflation ceases. However, the decay of the false vacuum does not occur at exactly the same instant throughout space. As in all quantum processes, there will be fluctuations in the rate at which the false vacuum decays. Some regions of the universe will decay slightly faster than others. In these regions inflation will end sooner. Irregularities will therefore appear in the final state. The hope is that these irregularities can act as seeds or centres for gravitational clumping that eventually lead to galaxies and galactic clusters. The theorists have been modelling the fluctuation mechanism mathematically, though with mixed success. Generally, the effect is too big, the computed irregularities too pronounced. But the models used are crude, and a more refined approach could prove successful. Although the theory is tentative at this stage, it is at least possible to see the sort of mechanism that could give rise to galaxies without the need for special initial conditions.

In Guth's original version of the inflationary scenario, the false vacuum decayed abruptly into the 'true' vacuum, the lowest energy vacuum state, which we identify with empty space today. The way in which this change occurred was regarded as similar to a phase transition such as from a gas to liquid, for example. Bubbles of true vacuum were envisaged as forming at random in the false vacuum, and then expanding at the speed of light to encompass greater and greater volumes of space. To enable the false vacuum to live long enough for inflation to work its magic, the two states were separated by an energy barrier through which the system was obliged to 'quantum tunnel', similar to the way described in Chapter 2 for electrons. This model suffered from a major shortcoming, however: all the energy released from the false vacuum was found to be concentrated in the bubble walls, and there was no mechanism to distribute it evenly through the interior of bubbles. When the bubbles collided and coalesced, the energy would end up in tangled sheets. The resulting universe would contain severe irregularities, and the work of inflation in achieving large-scale uniformity would be ruined.

Improved versions of the inflationary scenario have been devised which circumvent these difficulties. In the new theory there is no tunnelling between the two vacuum states, but instead the parameters are chosen so that the decay of the false vacuum is very slow, giving the universe time enough to inflate. When decay eventually occurs the energy of the false vacuum is released throughout the 'bubble', which quickly heats up to 10^{27} K. It is assumed that the entire observable universe is contained within a single bubble. Thus, on an ultra-large scale the universe may be very irregular, but our own region (and much more beyond) lies within a domain of quiescent uniformity.

Curiously, Guth's original reason for inventing the inflationary scenario was to address an altogether different cosmological problem, namely the absence of magnetic monopoles. As explained in Chapter 9, the standard big bang theory predicts that a superabundance of monopoles would have been created in the primeval phase. It also happens that these monopoles are likely to be accompanied by other bizarre objects known as 'strings' and 'sheets' which are their one- and two-dimensional analogues. The problem was how to rid the universe of these undesirable entities. Inflation solves the monopole and related problems automatically because the enormous swelling of space effectively dilutes them to zero density.

Though the inflationary scenario remains a partially developed and speculative theory, it has thrown up a set of ideas that promise to change for ever the face of cosmology. Not only can we now contemplate an explanation for why there was a big bang, but we can begin to understand why it was as big as it was, and why it took the form that it did. We can start to see how it is that the large-scale uniformity of the universe has come about at the same time as the controlled smaller-scale irregularities such as galaxies. The primeval explosion that produced what we know as the universe need no longer be regarded as a mystery for ever beyond the scope of physical science.

The self-creating universe

In spite of the great success of inflation in explaining the origin of the universe, a mystery remains. How did the universe arrive in the false vacuum state in the first place? What happened *before* inflation?

A completely satisfactory scientific account of the creation would have to explain how space (strictly spacetime) came to exist, in order that it might then undergo inflation. Some scientists are content to assume either that space always existed, or that its creation lies beyond the scope of science. A few are more ambitious, however, and believe that it is possible to discuss how space in general, and the false vacuum in particular, might have come out of literally nothing as a result of physical processes that are in principle amenable to study.

As already remarked, the belief that nothing can come out of nothing has only recently been challenged. The cosmic bootstrap comes close to the theological concept of creation *ex nihilo*. It is certainly true that in the familiar world of experience objects usually owe their existence to other objects. The Earth was formed from the solar nebula, the solar nebula from the galactic gases, and so on. If we should happen to encounter an object suddenly appearing out of nowhere we should be inclined to regard the event as a miracle: imagine locking an empty safe and then opening it a few moments later to find it full of coins, or cutlery, or candies? In daily life we expect that everything has to come from somewhere, or out of something.

On the other hand, the situation is not so clear cut in the case of less concrete things. Out of what is a painting, for example, created? Brushes and paints and canvas are, of course, needed, but these are tools. The *form* of the painting – the choice of shapes and colours, the texture, the composition – is not created by the paint and brush. It is the result of ideas.

Are thoughts and ideas created out of something? Thoughts surely exist, and perhaps all thoughts need a brain, but the brain is the mode of realization of the thoughts, not their cause. Brains alone do not create thoughts any more than computers create calculations. Thoughts can be created by other thoughts, but that still leaves the origin of thoughts unexplained. Sensations lead to some thoughts; memory also produces thoughts. Most artists, however, would regard their work as a result of *spontaneous* inspiration. If this is so, creating a painting – or at least the idea of a painting – is a form of creation out of nothing.

Nevertheless, can we conceive of physical objects, or even the entire universe, coming into existence out of nothing? One place where such a bold possibility is taken seriously is on the east coast of the United States where there is a curious concentration of theoretical physicists and cosmologists who have been manipulating mathematics in an attempt to divine the truth about creation *ex nihilo*. Among this esoteric coterie is Alan Guth at MIT, Sidney Coleman of Harvard, Alex Vilenkin of Tufts University, and Ed Tryon and Heinz Pagels in New York. All of them believe that in one sense or another 'nothing is unstable' and that the physical universe blossomed forth spontaneously out of nothing, driven by the laws of physics. 'Such ideas are speculation squared', concedes Guth, 'but on some level they are probably right . . . It is sometimes said there is no such thing as a free lunch. The universe, however, is a free lunch.'

In all these conjectures it is the quantum factor that provides the key. The central feature of quantum physics, as we saw in Chapter 2, is the disintegration of the cause–effect link. In the old classical physics, the science of mechanics exemplified the rigid control of causality. The activity of every particle, each twist and turn, was considered to be legislated in detail by the laws of motion. A body was understood to move continuously in a well-defined way according to the pattern of forces acting upon it. The laws of motion embodied the link between cause and effect in their very definition, so that the entire universe was supposed to be regulated in every minute respect by the existing pattern of activity, like a gigantic clockwork. It was this all-embracing, utterly dependable causality that prompted Pierre Laplace's claim about a powerful calculator being able to compute the entire history and destiny of the cosmos from the operation of mechanical laws. The universe, according to this view, is for ever unfolding along a pre-ordained pathway.

Quantum physics wrecked the orderly, yet sterile Laplacian scheme. Physicists learned that at the atomic level matter and motion are vague and unpredictable. Particles can behave erratically, rebelling against rigidly prescribed motions, turning up in unexpected places without discernible reason and even appearing or disappearing without warning.

Causality is not completely absent in the quantum realm, but it is faltering and ambiguous. If an atom, for example, is excited somehow by a collision with another atom, it will usually return quickly to its lowest energy state by emitting a photon. The coming-into-being of the photon is, naturally, a consequence of the atom's being excited in the first place. We can certainly say that the excitation caused the creation of the photon. In that sense cause and effect remain linked. Nevertheless, the actual moment of creation of the photon is unpredictable; the atom might decay at any instant. Physicists can compute the expected, or average, delay before the photon appears, but they can never know in any individual case when this event will happen. Perhaps it is better to describe such a state of affairs by saying that the excitation of the atom 'prompts' rather than causes the photon to come into being.

The quantum microworld is not, therefore, linked by a tight network of causal influences, but more by a pandemonium of loosely obeyed commands and suggestions. In the old Newtonian scheme a force would address a body with the unchallengable imperative 'You will move!' In quantum physics the communication is more of an invitation than an order.

Why do we find the idea of an object abruptly appearing from nothing so incredible? What is it about such an occurrence that suggests miracles and the supernatural? Perhaps the answer lies with familiarity. We never encounter the uncaused appearance of objects in daily life. When the conjurer pulls the rabbit out of a hat we know we have been duped.

Suppose that we actually lived in a world where objects did from time to time noticeably pop out of nowhere, for no reason, in a completely unpredictable way. Once we had grown accustomed to such events we would cease to marvel at them. Spontaneous creation would be accepted as a quirk of nature. Maybe in such a world it would no longer strain credulity to imagine the entire physical universe bursting into existence from nothing.

The imaginary world described above is not, in fact, so very different from the real world. If we could actually observe the behaviour of atoms directly with our sense organs, rather than through the intermediary of special instruments, we should frequently see objects appearing and disappearing without well-defined reasons.

The closest known instance to the idea of creation out of nothing occurs if an electric field can be made strong enough. At a critical field strength, electrons and positrons start appearing out of nowhere in an entirely random way. Calculations suggest that near the surface of a uranium nucleus the electric field is intense enough to be on the verge of inducing this effect. If nuclei could be made containing about 200 protons (uranium has 92) then the spontaneous creation of electrons and positrons would be observed. Unfortunately, a nucleus with so many protons is likely to be exceedingly unstable, but nobody is sure about this.

The spontaneous creation of electrons and positrons in an intense electric field can be considered as a bizarre type of radioactivity, in which it is empty space – the vacuum – which decays. We have already encountered the idea of one vacuum state decaying into another. Here the vacuum decays into a state containing particles.

Although the decay of space is difficult to achieve using an electric field, an analogous process involving gravity might well occur naturally. Near the surface of black holes, gravity is so intense that the vacuum sizzles with a continual stream of newly created particles. This is the famous black hole radiation discovered by Stephen Hawking. Gravity is ultimately responsible for creating the radiation, but it does not cause it in the old Newtonian sense: no given particle has to appear at any particular place and time as a result of gravitational forces. In any case, gravity is only a warping of spacetime, so we could say that it is spacetime that induces the creation of matter.

The spontaneous appearance of matter out of empty space is often referred to as creation 'out of nothing', and comes close to the spirit of the creation *ex nihilo* of Christian doctrine. For the physicist, however, empty space is a far cry from nothing: it is very much part of the physical universe. If we want to answer the ultimate question of how the universe came into existence it is not sufficient to assume that empty space was there at the outset. We have to explain where space itself came from. The idea of *space* being created might seem exotic, yet in a sense it is happening around us all the time. The expansion of the universe is nothing but a continual swelling of space. Every day the region of the universe accessible to our telescopes swells by 10^{18} cubic light-years. Where is all this space 'coming from'? A helpful analogy is with a piece of elastic. When an elastic string is stretched you get 'more of it'. Space is rather like super-elastic in that it can go on stretching for ever (as far as we know) without 'snapping'.

The stretching and warping of space also resembles elastic inasmuch as the 'motion' of space is subject to laws of mechanics in the same way as matter. These are the laws of gravity. Just as the quantum theory applies to the activities of matter, so it applies to space and time. In earlier chapters we have seen how quantum gravity is an indispensable part of the search for the superforce, which suggests a curious possibility: if quantum theory allows particles of matter to pop into existence out of nowhere, could it also, when applied to gravity, allow space to come into existence out of nothing? And if so, should the spontaneous appearance of the universe 18 000 million years ago occasion such surprise after all?

Free lunch?

The concept of quantum cosmology – applying the quantum theory to the entire universe of spacetime and matter – is being taken ever more seriously by theorists. Superficially, quantum cosmology seems a contradiction in terms. Quantum physics deals with the smallest systems, while cosmology is the study of the largest. Nevertheless, the universe was once very shrunken, and there must have been a time when quantum effects were important. Calculations suggest that quantum physics cannot be ignored at the GUT era (10^{-32} s) and would probably have dominated everything at the Planck era (10^{-43} s). It was at some moment between these two epochs when, according to theorists such as Vilenkin, the quantum universe erupted into existence. In the words of Sidney Coleman, 'We make a quantum leap from nothing into time.' Spacetime, it seems, is a fossil from this era.

Coleman's 'quantum leap' could be described as a form of 'tunnelling'. We have seen how in the original inflation theory the false vacuum state was required to tunnel into the true vacuum state through an energy barrier. In the case of the spontaneous appearance of the quantum universe out of nothing, however, our intuition is strained to the limit. One end of the 'tunnel' represents the physical universe of space and time, which 'arrived' by quantum tunnelling from nothing, and so the other end of the tunnel must be 'nothing'! Perhaps it would be better to say that there is only one end to the tunnel; the other end does not exist.

A central challenge facing these attempts to explain the origin of the universe is to account for the fact that the cosmos was created in a false vacuum state. Had the newly created spacetime been in the true vacuum, inflation would never have occurred, the big bang would have been reduced to a whimper, and spacetime would have shrunk back out of existence after a fleeting instant, devoured by whatever quantum activity produced it initially. Without being in the false vacuum the universe could never have locked into the cosmic bootstrap to make concrete its ephemeral existence. It may be that a false vacuum state was favoured by the extreme conditions prevailing at the time. For example, if the universe was created at a high enough initial temperature and then cooled, it might indeed have been stranded in a false vacuum. At the time of writing many technical questions of this sort remain unresolved.

Whatever the truth about these deep conceptual issues, the universe must have come into existence somehow, and quantum physics offers the only branch of science in which the concept of an event without a cause makes sense. When the subject at issue concerns spacetime it is in any case meaningless to talk about a cause in the usual sense. Causation is rooted in the notion of time, and so any ideas about an agency creating time, or causing time to come into existence, must appeal to a wider concept of causality than is at present familiar in science.

If space really is ten-dimensional, then the theory suggests that in the very early stages all ten dimensions enjoyed equal status. An attractive possibility is that the spontaneous 'compactification' – or rolling-up – of seven dimensions could be connected with the phenomenon of inflation.

According to this scenario the driving force for inflation is generated as a by-product of the forces which manifest themselves through the extra space dimensions. The ten-dimensional space might then evolve naturally in such a way that three space dimensions embark upon inflation at the expense of the other seven, which shrink away to invisibility. Thus, a microscopic quantum blob of ten-dimensional space suffers a spasm which inflates three dimensions to form a universe, and traps the remaining seven in a permanent microcosmos from which they are manifested only indirectly, as the forces of nature. It is an appealing theory.

Although much theoretical work remains to be done on the physics of the very early universe, it is possible to give a broad outline of the events which have shaped the cosmos as we see it today. In the beginning the universe erupted spontaneously out of nothing. From a featureless ferment of quantum energy, bubbles of empty space began to inflate at accelerating rate, bootstrapping colossal reserves of energy into existence. This false vacuum, infused with self-created energy, was unstable and began to decay, dumping its energy in the form of heat, filling each bubble with a fireball. Inflation ceased, but the big bang was started. The time was 10^{-32} s.

Out of the fireball came all matter and all physical structures. As the fireball cooled so the cosmic material suffered a sequence of phase transitions. At each transition more and more structure 'froze out' of the primeval stuff. One by one the forces of nature separated themselves. Step by step the objects we now call subatomic particles acquired the labels of their present identity. As the soup of matter became more and more complex, on a larger scale the irregularities left over from the inflationary phase began to grow into galaxies. With the further structuring and specialization of matter, the universe began to acquire a more recognizable form, with hot plasma condensing into atoms, forming stars, planets, and eventually life. Thus, the universe became self-aware.

Matter, energy, space, time, forces, fields, order, and structure: these are the items on the Creator's shopping list, the indispensable requirements for a universe. The new physics holds out a tantalizing promise that we might explain from science how *all* these things came to exist. No longer do we need to 'put them in by hand' at the beginning. We can see how all the fundamental features of the physical world could have arisen *automatically*, purely as a consequence of the laws of physics, without the need to assume that the universe was set up in a very special state initially. The new cosmology tells us that the initial cosmic state is irrelevant, all information about it is destroyed during the inflationary phase. The universe we see bears only the imprints of the physical processes which have happened since the onset of inflation.

For millenia mankind has believed that nothing can come out of nothing. Today we can argue that everything has come out of nothing. Nobody needs to pay for the universe. It is the ultimate free lunch.

13
The Unity of the Universe

All things by immortal power
Near or far
Hiddenly, to each other linked are
That thou canst not stir a flower
Without troubling of a star.

Francis Thompson (1859–1907)

The concept of a universe

The word *universe* has the same origin as *unity* and *one*. It means, literally, the totality of things considered as a whole. Curiously, the word *wholly* derives from the same origin as *holy*, which reflects the deep mystical and metaphysical associations of cosmology. Indeed, until the twentieth century, the study of the universe as a whole lay almost exclusively in the province of religion. Scientific cosmology is only a very recent subject.

The mystical appeal of cosmology accounts for its considerable popularity both among scientists and the public at large. Indeed, many people draw scant distinction between scientific cosmology, mysticism, and blatantly crackpot topics on the outer fringes of the so-called paranormal. In spite of this muddle, I believe that the widespread interest in cosmology is a good thing in a world where fragmentation and conflict frequently triumph over unity.

Talk of 'the universe' has now become so commonplace that it has obscured what is perhaps the most remarkable of all cosmological facts, and that is that the concept of universe is meaningful at all. How is it possible that we can treat all physical existence as a whole?

There is a deep philosophical issue involved in this question. Science is founded upon the concepts of laws and experimental tests. A scientific theory is a coherent account of some aspect of nature based on a collection of consistent principles, preferably expressed in mathematical form. The theory is intended to be a model of part of the world. It stands or falls on its utility. Other scientists are invited to carry out experiments to determine how closely the model fits reality. If these experiments confirm the accuracy of the model, repeatedly, confidence in the theory grows and it becomes part of the accepted body of science until such time as a better, more accurate, or more comprehensive theory is developed.

An essential part of the scientific method rests with the repeatability of experimental tests. To take a simple example, it was claimed by Galileo that all bodies accelerate at equal rates when falling, so that two bodies when dropped together will strike the ground together, even if they are of different weight. This claim was greeted with general scepticism because for centuries people had believed the Aristotelean doctrine that heavy objects fall faster, an idea that perhaps accords better with intuition. Whatever the beliefs, it is an easy enough matter for Galileo's claim to be tested, by the simple expedient of dropping a few things to find out. When this had been done often enough, people came to accept Galileo's version of the way that material bodies fall.

In the example chosen, testing the theory is easy because of the availability of a limitless supply of small bodies for dropping. The situation in cosmology, however, is completely different. By definition there is only one universe. There is no question of having a 'cosmic law' because such a law could never be tested by repeating experiments on a collection of similar systems. This raises the profound question of how we can apply scientific reasoning at all to the universe as a whole.

In practice, cosmologists fall back on the idea of extrapolation. The laws of physics which are deduced from experiment and observation on parts of the universe are simply taken, unaltered, and applied to the totality. Thus, the general theory of relativity (our best current model of gravity), which is tested chiefly by observations within the solar system, is nevertheless used to compute the motion of the whole universe. Remarkably enough, this procedure seems to work. Using the laws that apply to a fragment of the universe to describe the whole cosmos seems to give a very plausible account of the observed state of affairs. Why?

Answering this question brings me back to the original problem of how we can talk meaningfully about 'the universe' in the first place. There is an analogy here with human society. A politician may reason as follows: 'I like tax cuts, my friends like tax cuts, people interviewed by opinion pollsters like tax cuts. Therefore, the country will be pleased with a tax cut.' The assumption here is that society as a whole has a sort of collective consciousness which reflects the predilections of its individual members. This enables the same principles that work for individuals to be applied to the totality. However, the reasoning works only if society is made up of reasonably like-minded individuals. On the question of tax cuts most people feel the same. The result might be very different if the issue concerned, say, religious practices.

In applying the laws of physics to the universe as a whole we make the same sort of logical leap as with the tax cuts. The universe is composed of lots (perhaps an infinity) of similar or identical systems. On a large scale we can think of the universe as a collection of galaxies, on a finer scale as a collection of atoms. At the deepest level the universe is a collection of quantum fields. The fact that right across the observable universe we see the same sorts of object is frequently taken for granted. Yet it is by no means obvious why this astonishing universality should exist.

The universality of physical systems is the starting point for scientific cosmology. A survey of the sky reveals that the stars are very similar to our sun, and other galaxies strongly resemble our own Milky Way in both size and structure. Closer analysis shows that these distant bodies are composed of the same atoms that we find on Earth. A terrestrial atom is indistinguishable from an atom on the very edge of the observable universe. The physical processes that occur in the most remote regions of the cosmos appear to be exactly the same as the processes that occur in our own cosmic neighbourhood. Most significantly, the forces of nature are universal. For example, the strength of the electromagnetic force in distant quasars can be inferred from a careful study of their light spectra. There is no noticeable difference from the electromagnetic force we observe in the laboratory.

As astronomers have expanded their horizons to encompass wider and wider regions of the universe, they have generally found more of the same. Why this should be so is not at all clear. Several centuries ago mankind believed that the Earth was the centre of creation, unique in form and location. Since Copernicus, all the evidence has pointed in the opposite direction, i.e. that the Earth is a typical planet, in a typical galaxy, situated in a typical region of the universe, and that the universe is made up of a superabundance of more or less similar things.

Scientists formalize these ideas into something called the 'cosmological principle' which, loosely speaking, says that our local neighbourhood of the universe is typical of the whole. This applies not only to atoms, stars, and galaxies, but to the general organization and distribution of both matter and energy. The universe is extraordinarily uniform, both in the way that the galaxies are scattered in depth throughout space and in their orientation around us. As far as we can see, there are no privileged places or directions in the cosmos. Moreover, this uniformity is preserved with time, as the universe expands; the expansion rate is the same in all regions of space and in all directions. Indeed, it is hard to envisage a universe much simpler in form that would be consistent with the existence of living observers. In earlier chapters we found a very persuasive reason for this large-scale cosmic co-operation in the so-called inflationary theory of the universe.

The scientific picture of the cosmos is, therefore, one of uniformity, coherence, and large-scale simplicity. If the universe were to be expanding at drastically different rates in different directions, or were to contain large variations in the density and arrangement of matter, it is doubtful whether there would even be a subject of scientific cosmology. (In fact, there probably would be no scientists either.) It is this uniformity, coherence, and simplicity which enables us to talk about 'the universe' as a single entity. Until very recently the origin of these qualities was a mystery. Now we can see that the instructions for building a coherent, uniform cosmos are written into the laws of physics. The superforce contains just the right features to take command of the early universe and organize it into a unified structure with the large-scale simplicity we now observe.

Mach's principle: linking the large and the small

Although we can all recognize the strong pervasive unity of *form* in the universe, there is a compulsive desire to search for a deeper cosmic unity, one which weaves together our own local region with the grand totality in some intimate way. Linking the large and the small, the global and the local, has a strong appeal because it makes us feel at one with all creation, a mystical objective common to most of the world's religions. Many people doubtless feel themselves to be linked spiritually to the totality of things, but there is also a parallel tradition in science for forging such links.

An early scientific argument for the existence of a deep connection between the large-scale structure of the universe and local physics was enunciated by the Austrian physicist and philosopher Ernst Mach (1838–1916), who has also achieved immortality from the use of 'Mach numbers' as a unit of the speed of sound. Although Mach espoused some erroneous ideas (he did not believe in atoms), his work on the nature of inertia, subsequently dignified with the title of Mach's principle, has proved one of science's most enduring speculations. Certainly, Mach's ideas exercised a profound influence on the young Einstein in his attempts to formulate the general theory of relativity. Einstein explicitly acknowledged his indebtedness to Mach in a letter written in June 1913, following the publication of Mach's book *The Science of Mechanics* the previous year.

Mach was born in the town of Turas in what is now Czechoslovakia. He held professorships in both mathematics and physics at the University of Graz, before moving to Prague, and then to Vienna as Professor of Philosophy, where he embraced the so-called positivist movement. Mach believed that reality must be rooted in observations, and it was this outlook that contributed to his cosmological ideas.

Mach was deeply interested in the nature of motion; in particular, he was concerned about the distinction that had been made between real and apparent motion. Our ancestors believed that the heavens rotated around the Earth, that the Earth was fixed at the centre of the universe, and that the sun, moon, and stars all moved along curved paths. It was a perfectly natural belief, because the heavenly bodies can be seen moving across the sky. By the seventeenth century such ideas had been discredited, however, and the movement of the heavenly bodies was seen to be only apparent. In reality it is the Earth that is rotating.

How should we prove to a sceptic that the rotation of the stars is only apparent and that it is the Earth that spins on its axis? A good way might be to appeal to Newton's science of dynamics. The rotating Earth experiences centrifugal effects, which cause it to bulge at the equator. Careful measurements of the Earth's geometry show that it is 43 km wider across the equator than from pole to pole. The reason that rotation causes equatorial bulging can be traced to the existence of inertia.

Inertia is a property of matter familiar to us all. Heavy objects have a high inertia, which means that they are hard to move, but once you get them moving they are hard to stop. Light objects can be moved about more easily. It is the inertia of the Earth that keeps it hurtling through space. Without its inertia, the Earth would stop in its orbit and fall into the sun. Inertia projects you out of your seat when a car brakes abruptly, and 'leaves your stomach behind' in a suddenly descending elevator. It is your inertia that tries to throw you off a rotating merry-go-round, or pins you to the wall of a spinning centrifuge. Inertia bursts the flywheel that is turning too fast, and it is this tendency for material to be 'thrown off' from rotating bodies – sometimes called centrifugal force – that is responsible for causing the Earth's equatorial bulge.

How should we relate the force of inertia to the other forces of nature? This is a puzzle that goes right back to Newton himself, and his first systematic description of the laws of motion. The central feature of Newton's work was his recognition that uniform motion – motion at a constant velocity – is entirely relative. Imagine that you are enclosed in an opaque box out in the depths of space. There is no way that you could tell whether the box was at rest or moving uniformly. This state of affairs is closely approximated on board an aircraft in level flight. Our experience of force and motion in the aircraft is indistinguishable from that in a room on the ground. The uniform motion of the aircraft in no way affects how objects within the aircraft behave; walking, eating, breathing, and all other activities appear normal.

Why, then, do we say that the aircraft is moving? To be sure, if we look out of the window we can see the ground passing by beneath, but what we really mean by motion here is that the aircraft is moving *relative* to the ground. The ground is not, of course, at rest. The Earth is in orbit around the sun (we don't feel that either) and the sun is moving around the galaxy.

An important consideration here is the fact that space itself has no landmarks, and so it is impossible to gauge our motion through space as such. One region of space looks exactly like another. There is no way you can see or feel space 'rushing past' as can a fish swimming through the ocean. There is no 'slipstream' to help us measure our speed. Being 'at rest' in space has no observational significance at all, a fact well appreciated by Newton: 'For it may be that there is no body really at rest to which the place and motions of others may be referred.'

The relativity of uniform motion is built into Newton's laws of mechanics, which state that no force or physical agency is required to maintain uniform motion. On the contrary, a body will continue to move uniformly unless something intervenes to alter this state of affairs. In the absence of external forces the body is continually carried forward by its inertia.

On Earth it is very hard to remove the interfering effect of forces. A puck propelled across a surface of ice comes close to free motion. The puck's inertia causes it to keep moving at a reasonably undiminished speed, without the need for any propulsive force, once it has been struck. In contrast, an automobile experiences so much friction and air resistance that its inertia is soon overcome by these forces. The automobile will come to rest in a short distance once the engine has been switched off.

In contrast to the relativity of uniform motion, accelerated or non-uniform motion has quite a different character. If an aircraft suddenly banks, dives, or increases power, the passengers are immediately aware of the disruption from the way they are thrown about. Even inside an opaque box in space, accelerated motion would be instantly spotted.

How is it that we recognize accelerated motion so readily? The key is inertia. Bodies subjected to acceleration try to resist in a noticeable way. Rotation is a particular case of accelerated or non-uniform motion. If the opaque box started to rotate, you would feel yourself pressed against the walls as your body tried to follow a straight path and the spinning box forced you along a curve. This seems to suggest that whereas uniform motion is relative to other bodies, accelerated motion is absolute.

Some scientists and philosophers have been unable to accept this conclusion. A near-contemporary of Newton, the Irish philosopher Bishop George Berkeley, wrote 'I believe we may find all the absolute motion we can frame an idea of, to be at bottom no other than relative motion'. Berkeley's argument was that, as space is featureless, we cannot conceive of *any* form of motion through space as such. Only by referring to other material bodies can we make sense of the idea:

> 'It suffices to replace "absolute space" by a relative space determined by the heaven of the fixed stars . . . Motion and rest defined by this relative space can be conveniently used instead of absolutes.'

Berkeley here introduces a crucial and intriguing factor, the 'fixed stars'. Today we know that the stars are not actually fixed, but are themselves in motion about the galaxy. Nevertheless, this motion is barely discernible because the stars are so far away. The important suggestion Berkeley makes is that the very distant matter in the universe in some way acts as a standard or reference frame against which we can judge all motion.

Behind this debate on motion lies the whole question of the nature of space and the distinction between space and nothingness. Aristotle proclaimed that 'nature abhors a vacuum' and argued that emptiness is nothingness and therefore cannot exist. The apparent space between bodies can be conceived of only by supposing it to be continuously filled with substance, ethereal or otherwise.

The parallel tradition of the void – empty space existing in its own right – also attracted its adherents. Among those was Newton, who discussed what he called 'Absolute space . . . without relation to anything external'. Newton's absolute space was ridiculed by his rival, Gottfried Leibniz, who declared, 'There is no space where there is no matter.'

Modern physics cuts right across this ancient controversy by replacing space with a quantum vacuum which supplies a sort of texture to what is superficially just emptiness. Yet the quantum vacuum, with its frolic of virtual particles, is a far cry from the continuous fluid envisaged by Aristotle.

Newton believed he could prove the existence of absolute space scientifically, by referring to inertial effects. The equatorial bulge of the rotating Earth demonstrates that it is the Earth which rotates, not the stars. The rotation of the Earth, Newton declared, is not merely relative to the stars, it is absolute. The Earth *really is* rotating in absolute space.

It was this last assertion that Berkeley challenged, claiming that if the universe were empty of all bodies save one, the concept of motion – uniform or otherwise – would be meaningless. 'If', wrote Berkeley, 'a globe were to exist alone, no motion could be conceived of it.' We should be unable to determine whether it was rotating or not. Berkeley went on, 'Let two globes be conceived to exist and nothing corporeal besides them.' In these circumstances we can make sense of the relative motion of the globes towards or away from each other, but still 'a circular motion of the two globes around a common centre cannot be conceived by the imagination'. On the other hand, 'Let us suppose that the sky of fixed stars is created; suddenly from the conception of the approach of the globes to the different parts of the sky the (revolutionary) motion will be conceived.'

Notwithstanding the success of Newtonian mechanics, these sentiments of Berkeley remained alive and were echoed two centuries later by Mach, who refused to draw a fundamental distinction between uniform and non-uniform relative motion, proclaiming that 'accelerated and inertial [i.e. uniform] motions result in the *same* way'. But how was Mach to reconcile his belief that even accelerative motions such as rotation are purely relative, with the existence of inertial forces, such as the centrifugal effects which cause a rotating body to bulge at the equator? After all, Newton had laid down an explicit challenge to those who doubted the existence of absolute motion: 'The effects which distinguish absolute from relative motion are centrifugal forces . . . For in a circular motion which is purely relative no such forces exist.'

The reasoning that Mach adopted to meet this challenge is boldly direct. If rotation is merely relative to the 'fixed' stars, he argued, then the centrifugal forces experienced by a rotating body must be *caused* by the stars. Mach's hypothesis amounts to nothing less than the claim that inertia has its origin in the far depths of the universe. If this explanation for the origin of inertia is accepted, Newton's absolute space can be discarded, and all motion treated as relative. It is a line of argument known today as Mach's principle, and it has exercised a strange fascination on several generations of physicists. It has also prompted some strong criticisms. Even Lenin felt moved to denounce the idea.

Can Mach's principle be made to work? The first problem is to explain the nature of the linkage which enables the far-flung stars to bestow inertial effects upon a body on Earth, or anywhere else in the universe. We get a clue from the fact that centrifugal force feels just like a gravitational force. One plan for a future space station is a wheel-shaped structure designed to rotate about its axis at the correct speed to simulate one '*g*' at its periphery. This is based on the idea of 'artificial gravity'. The close similarity between centrifugal and gravitational forces was well understood by both Galileo and Einstein. Indeed, it is a founding principle of the general theory of relativity that, locally, the two forces *are* identical. It is therefore natural to look to the gravitational field of the universe to explain centrifugal and other inertial forces.

How could the gravity of the stars cause inertia? A possible idea might be to suppose that a rotating body sends out some sort of gravitational influence which is received by the stars. The stars are disturbed slightly, and as a result generate their own gravitational effect that reacts back on the rotating body. The reaction produces what we call a centrifugal force, but this force is actually a gravitational effect of cosmic origin. Admittedly, the contribution of any given star to the centrifugal force must be exceedingly small due to the great distances involved, but the number of stars is so numerous that the cumulative effect could still be significant. It is a fascinating conjecture. Every time you 'leave your stomach behind' it is the remote galaxies, thousands of millions of light-years away, that are pulling at it!

The trouble with the simple picture given above is that, according to the theory of relativity, a gravitational disturbance cannot travel faster than light. Even at the speed of light it would take many millions of years for the gravitational 'echo' of a rotating body to come back. But we know that centrifugal effects occur instantaneously, as soon as a body starts to rotate.

Einstein believed that he had found a way to overcome the time delay problem by formulating Mach's principle as part of his cosmological investigations. Strangely, he could only get the scheme to work if the universe is curved. And not only curved, but with the sort of curvature that causes it to be spatially closed (a hypersphere). Infinite, unbounded space would not do. There followed a protracted and confused debate, which continues to this day, as to what extent Einstein's general theory of relativity does or does not incorporate Mach's principle.

In 1949 the mathematician and logician Kurt Gödel discovered a solution of Einstein's gravitational field equations which actually describes a rotating universe. Gödel's model is not intended necessarily to represent the real universe, but it is nevertheless a logical possibility within Einstein's theory. According to Mach's principle, a rotating universe is an impossible concept, for relative to what is the whole universe rotating?

On the other hand, some effects predicted by general relativity have a distinctly Machian flavour. One of these was mentioned by Einstein in his letter to Mach. Suppose we accept that the inertial forces on a body are due to the gravitational action of all the other matter in the universe. The predominant effect will obviously arise from matter at a great distance, because that is where most of it is located. Nevertheless, nearby bodies must still exert a tiny effect. Einstein invited Mach to consider a body enclosed within a heavy shell of material, which shell is set rotating relative to the fixed stars. If Mach's ideas are right the reference frame against which the motion of the incarcerated body is to be gauged is some sort of average associated with all the other matter in the universe, and in taking this average the spherical shell cannot be excluded. Its contribution to the overall cosmic reference frame will certainly be very small, but it will not be zero. Its value can be computed from theory. Calculation shows that, as Mach would have expected, the rotation of the shell does indeed generate a tiny inertial force. This force acts on the body inside the shell and tries to make it co-rotate.

Remarkably, such effects may actually be observable. Consider, for example, the experiences of a gyroscope in orbit about the Earth. The motion of the gyroscope is affected by the Earth's spacewarp, and as the Earth rotates so it 'twists' the spacewarp. This has the effect of twisting the gyroscope too. The effect is very small – it would take millions of years for the gyroscope to twist around once – but nevertheless the twisting motion could probably be detected with current technology by putting the gyroscope inside a protective casing to remove other, non-gravitational disturbances such as the solar wind. A project of this sort has been planned for several years by Professor William Fairbank of the University of Stanford, the same man who was mentioned in Chapter 8 in connection with the free quarks experiment.

In more massive bodies the 'dragging' effect of a rotating spacewarp can become pronounced. The most extreme case is that of a rotating black hole, where a nearby object can be dragged around with such violence that no force in the universe can stop it. This spectacular action is sometimes referred to in the popular literature as a 'space vortex' surrounding the hole.

It is probable that Mach's principle cannot be verified experimentally. How could we ever know whether a rotating Earth would bulge in an otherwise empty universe when there is no way that we can remove the rest of the universe to find out? On the other hand, it is possible to imagine an experiment by which the principle could be falsified. If it were determined, on the basis of very careful measurements, that the entire universe is rotating in an absolute sense, then Mach's principle would be discredited.

An absolute cosmic rotation would single out a preferred axis in space, and we might expect this privileged direction to manifest itself in the arrangement of matter and energy in the universe. The cosmic heat radiation is known to be uniform in all directions to at least one part in ten thousand, from which a stringent limit can be set on any cosmic rotation that might exist. In fact, it can be shown that the universe could not have revolved by more than a few degrees in its entire history. Thus, to a high degree of accuracy at least, the motion of the universe seems to be consistent with Mach's principle.

Signals from the future

Like many professional scientists I am a science fiction addict. A couple of years ago I sat down to read Gregory Benford's *Timescape*. Imagine my surprise when, fairly early in the plot, I came across a character called Paul Davies, a physicist with a passionate interest in time, who declared with authority that it might be possible to send signals into the past. His advice was duly taken, and the hero of the plot set about attempting to communicate with a scientist of a previous generation in order to save the world from catastrophe.

My unexpected fictionalization was due to a longstanding interest of mine in the nature of time. I first became fascinated by the idea of signalling the past after listening to a lecture by Fred Hoyle given at the Royal Society in London when I was a student. Hoyle pointed out that Maxwell's famous electromagnetic field equations, which describe the propagation of electromagnetic waves, already contain within them the possibility that such waves could travel backwards in time.

This startling conclusion can be understood by analogy with ordinary water waves. If you throw a stone into a still pond, the ripples created travel outwards from the point of disturbance and fade away at the pond's edges. Such outgoing wave patterns are easy to make. On the other hand, we never encounter organized wave patterns appearing at the edges of a pond and converging to a point. However, the physical processes which control the undulations are perfectly reversible. Each portion of the wave could be made to run backwards. In spite of this, only the outgoing type of waves are produced spontaneously in nature. True, one can generate converging waves artificially – say by dropping a ring horizontally on to the pond's surface – but this is much harder to achieve than the production of outgoing waves. Why?

The one-way character of wave disturbances extends to all types of wave motion and imprints on our universe an 'arrow of time', i.e. a distinction between past and future. If you were to film the ripples on a pond and play the film backwards, the deception would be immediately obvious. In the case of electromagnetic waves, such as radio waves, the idea of a coherent pattern of waves converging to one place seems preposterous. As radio waves can propagate to the ends of the universe, the only way a converging pattern could be produced is by some huge cosmic conspiracy in which waves come in from infinite space in all directions, precisely in step.

Because of the connection between wave motion and the arrow of time, we can think of outgoing waves as travelling into the future in the usual way, but incoming or converging waves as time-reversed, i.e. travelling into the past. The former are called 'retarded' waves, because they arrive after they are sent, whereas the latter are known as 'advanced', for they arrive in advance of their transmission. Ever since Maxwell, it has been believed that advanced electromagnetic waves are *possible* – they are logically permitted by the theory – but are as physically ridiculous as time travel, and so must be discarded.

Most scientists have been happy enough to reject advanced waves as irrelevant, without questioning why the universe is set up to exclude them so thoroughly. Notable exceptions were John Wheeler and Richard Feynman. At the end of the Second World War they published an intriguing paper which attempted to demonstrate why retarded electromagnetic waves are the norm, and to explore the possibility that advanced waves (waves from the future) might exist. Wheeler was then a nuclear physicist who had worked with Niels Bohr and Enrico Fermi on fission, while Feynman was a student who, shortly afterwards, succeeded in formulating QED and winning a Nobel prize.

Wheeler and Feynman decided to investigate what would happen in a world where advanced and retarded waves exist on an equal footing. In such a hypothetical universe a radio transmitter would send signals equally into the past and the future. It might be thought that these circumstances would be bound to lead to nonsensical conclusions, but in a remarkable argument Wheeler and Feynman demonstrated that this need not be so.

Consider the fate of the troublesome advanced waves that leave the transmitter and travel out into space backwards in time. Eventually these waves will encounter matter, in the form of electrically charged particles, perhaps as tenuous gas in intergalactic space. The waves will set these charges into motion with the result that they too generate secondary waves of an identical frequency, also one-half retarded and one-half advanced in nature. The retarded portion of these secondary waves would then travel forward in time, thus creating a tiny echo at the transmitter at the instant of the original transmission. We thus get a complicated network of disturbances and echoes bouncing to and fro around the universe, both backwards and forwards in time.

Although the echo of any individual charged particle would be inconceivably small because of its great distance from the transmitter, if the universe were furnished with so many particles that it was effectively opaque to electromagnetic radiation, the cumulative effect of all the echoes would be precisely equal in strength to the original signal. Closer analysis shows something even more extraordinary. The echo, which interpenetrates the original advanced wave everywhere in space, turns out to be exactly out of phase with it. This has the effect of cancelling the advance wave completely by destructive interference. All signals sent into the past are precisely killed by their own echoes! Wheeler and Feynman therefore concluded that, in an opaque universe, only retarded electromagnetic waves will occur, even if every individual charged particle radiates symmetrically both advanced and retarded waves.

The astonishing result of the Wheeler–Feynman analysis comes about because, in their theory, the electromagnetic activity of any individual charged particle cannot be separated from that of the entire universe. The waves produced in one place cannot be untangled from the echoes they induce even from the most distant regions of the cosmos. Moreover, because of the ability of advanced signals to propagate backwards in time, there is no thousand-million-year delay for the echo to return (this was the problem about a gravitational 'echo' needed to produce a force of inertia). Therefore every humble radio transmission becomes literally a cosmic event.

The whole and its parts

The Wheeler–Feynman theory is Machian in the sense that it seeks to link the local and the global in a network of influences, and suggests that we may understand individual physical systems only by proper reference to the whole. Although the theory remains speculative, there is general agreement that the absence of advanced waves in nature ultimately requires a cosmological explanation, and that the arrow of time is of cosmological origin. And so the fact that we are aware of a sharp distinction between past and future in the behaviour of the world about us is an example of the linkage between the large and the small, the whole and its parts.

There may well be other linkages of this sort. The magnetic monopole provides another possible example. As explained in Chapter 9, when Dirac originally investigated the idea of magnetic monopoles, he found that the value of the magnetic charge carried on any monopole that might exist is tied through the laws of quantum electrodynamics to the fundamental unit of electric charge carried by the electron. One implication is that, even if there is only one magnetic monopole in the whole universe, it forces the electric charge on every electron to be what it is. Therefore the quantity of charge on an electron here may depend on the existence of a magnetic monopole on the other side of the cosmos.

In recent years much attention has been given to the role of quantum physics in providing a link between the part and the whole. An eloquent exposition has been given by David Bohm in his book *Wholeness and the Implicate Order*, in which he writes 'The quantum theory has a fundamentally new kind of non-local relationship, which may be described as a non-causal connection of elements that are distant from each other.'

Bohm draws an analogy between order in the quantum universe and order in a hologram. A hologram is a device for encoding information about a picture. The encoded scene can be reconstructed as a projected three-dimensional image using laser light. The information about the scene is stored as a pattern on a photographic plate, but in a form that cannot be recognized by the human eye. The pattern is in fact produced by interference between two laser beams, and is generally very complex. It can only be unscrambled by resorting to lasers once more. In a conventional photographic slide each feature of the projected image has its counterpart at one place on the slide: there is a one-to-one correspondence between parts of the image and parts of the slide. A hologram is quite unlike this. Every feature of a hologram image is encoded across the whole photographic plate. The difference is most apparent if only a fragment of the plate is illuminated. The re-created image remains intact, though somewhat degraded in quality, because the information for the whole image is still available even on a portion of the plate. This is in sharp contrast to a conventional slide, where incomplete illumination would give rise to a projected image with parts missing.

Other writers, such as Fritjof Capra in *The Tao of Physics* and Gary Zukav in *The Dancing Wu Li Masters*, have emphasized the close parallels between quantum physics and oriental mysticism, with its emphasis on the unity of existence and the subtle relationships between the whole and its parts.

The holistic world-view suggested by quantum physics is largely prompted by the non-local character of quantum states, as outlined in Chapter 3. It will be recalled that in the experiment of Einstein, Podolsky, and Rosen, two particles remain intimately linked although far apart. In such a situation it is not possible to consider each particle as existing separately, in a well-defined condition, in the absence of the other.

More generally, it is possible to regard a quantum particle as possessing a well-defined attribute, such as a position or a motion, only within the context of some specified experimental arrangement in which the apparatus is designed to measure a particular property of the particle. It makes sense, for example, to talk about a particle-at-a-place only if it is part of a complex system designed to measure its position. In the absence of a measurement context, all talk of the particle's location is meaningless. Thus, we can define the position of a quantum particle only within the framework of a macroscopic measuring system that itself contains countless millions upon millions of other quantum particles. The particle's position is really a collective or holistic concept.

There is clearly a subtle linkage here between the reality of the microscopic world and the familiar, large-scale, macroscopic world. Ultimately, we cannot separate the quantum reality from the structure of the entire universe, and so the state of an individual particle is meaningful only when it is regarded in the context of the whole. The microscopic and macroscopic worlds are intertwined and can never be separated.

The idea that non-causal, holistic order exists in the universe by no means originated with modern physics. Astrology, for example, is an attempt to discern a cosmic order in which the affairs of human beings are reflected in the organization of the heavens. The psychoanalyst Carl Jung and the quantum physicist Wolfgang Pauli proposed a non-causal connecting principle which they called synchronicity. They compiled evidence for a sort of pervasive order in which apparently independent events occur in conjunction in a meaningful way. Typical of such events are documented instances of extraordinary coincidences, well beyond the expectations of chance. A popular account of these ideas has been given by Arthur Koestler in his book *The Roots of Coincidence.*

An element of paradox runs through these ideas, reminiscent of Zen, and also of the 'strange loops' discussed by Douglas Hofstadter in his *Gödel, Escher, Bach*. The whole supports the parts which themselves constitute the whole. We need the universe before we can give concrete reality to the very atoms that make up the universe! Which 'comes first', atoms or universe? The answer is 'neither'. The large and the small, the global and the local, the cosmic and the atomic, are mutually supportive and inseparable aspects of reality. You can't have one without the other. The tidy old reductionist idea of a universe which is simply the sum of its parts is completely discredited by the new physics. There is a unity to the universe, and one which goes far deeper than a mere expression of uniformity. It is a unity which says that without everything you can have nothing.

14
A Cosmic Plan?

A rational universe

Steven Weinberg once wrote, 'The more the universe seems comprehensible, the more it also seems pointless.' Weinberg is one of the world's leading theoretical physicists and has probably done more than anyone else of his generation to unify physics. Co-architect of the unified theory of the weak and electromagnetic force, Weinberg is able to survey much of modern physics and cosmology with a uniquely expert gaze and to draw a particularly well-considered conclusion. His remark is typical of many made by scientists today, who infer from their research that the universe has no discernible purpose, and must be considered as a vast and meaningless accident.

Curiously, other scientists, surveying the same set of principles and the same technical data, arrive at quite different conclusions. Some, like Erwin Schrödinger, admit bewilderment: 'I know not whence I came, nor whither I go, nor who I am.' For them, nature is too subtle, too profound. We can scratch the surface of reality, but we will always leave beneath vast, unfathomable depths of mystery. All we can hope to do is to probe a few of the principles which administer the cosmos and marvel at the beauty of it all. Our scope of vision is far too narrow for us to grapple with deep issues of meaning and purpose.

A few scientists, however, are more bold, more positive. They readily concede that our knowledge of the workings of nature is limited and tentative, but they are optimistic that we shall eventually succeed in discovering the truly fundamental laws that govern the universe. John Wheeler has written, 'Some day a door will surely open and expose the glittering central mechanism of the world in its beauty and simplicity.'

There are even those who are prepared to suggest that the 'glittering central mechanism' may even now be within our grasp. Under the title 'Is the end in sight for theoretical physics?', Stephen Hawking delivered a lecture on the occasion of his inauguration to the Lucasian Chair at the University of Cambridge, a post formerly held by Newton. Hawking argued that supergravity provides, for the first time, the possibility of a unified theory of nature in which all physical structures and all processes are described by a single mathematical principle. Such an achievement would, he reasoned, represent the culmination of physical science. One could believe that the theory attained would not be merely yet another approximation on the endless road to truth, but the truth itself. We could then have the same conviction in this ultimate law of nature as we do in the rules of arithmetic.

Few physicists are prepared to go this far, but many have been deeply inspired by the remarkable harmony, order, and unity of nature that recent advances have uncovered. They are sufficiently impressed by the way that the laws of nature hang together that they feel compelled to believe there is something behind it all; in a pithy phrase of Fred Hoyle, 'The universe is a put-up job.'

What leads scientists to reach such powerful conclusions? In the previous chapter some of the evidence was given for a belief in an all-pervasive unity in nature. The study of cosmology offers especially compelling evidence for unity; the subject itself would not exist if one could not speak of 'the universe' as an integrated system.

But the evidence goes beyond unity. Every advance in fundamental physics seems to uncover yet another facet of *order*. The very success of the scientific method depends upon the fact that the physical world operates according to rational principles which can therefore be discerned through rational enquiry. Logically, the universe does not have to be this way. We could conceive of a cosmos where chaos reigns. In place of the orderly and regimented behaviour of matter and energy one would have arbitrary and haphazard activity. Stable structures like atoms or people or stars could not exist. The real world is not this way. It is ordered and complex. Is that not itself an astonishing fact at which to marvel?

How, therefore, do some scientists such as Weinberg conclude that the world is pointless in the face of the ubiquitous order displayed by the laws of nature? I think that in part it is a case of being unable to see the wood for the trees. A professional scientist is so immersed in unravelling the laws of nature that he forgets how remarkable it is that there are these laws in the first place. Because science presupposes rational laws, the scientist rarely stops to think about why these laws exist. Just as a crossword addict assumes without thinking that there is an answer to the puzzle, so the scientist rarely questions the fact that there are rational answers to his scientific enquiries.

This spirit of 'so what?' has permeated the whole of Western technological society. Even non-scientists accept without thinking the orderly operation of the cosmos. They know the sun will rise on schedule each morning, that a stone will unfailingly fall down rather than up, and that the gadgets and machines around them will always operate correctly so long as there are no faults in their mechanisms. The rationality, dependability, and order of the physical world is taken for granted. It is so much a fact of life it rarely provokes any sense of wonder.

The harmony of nature

In addition to its unity and order, physicists are deeply impressed by the unexpected harmony and coherence of nature. Traditionally, physics has been divided up into a number of rather distinct branches, such as mechanics, optics, electromagnetism, gravity, thermodynamics, atomic and nuclear physics, solid state, and so on. These rather artificial divisions conceal the elegance with which these topics dovetail together. We don't find, for example, that the laws of gravity conflict with those of electromagnetism or solid state physics. In many cases this consistency is not at all manifest, and is revealed only by careful analysis. A beautiful example close to my own area of research concerns the second law of thermodynamics. Although this law was originally deduced in the mid-nineteenth century for a rather limited range of processes having to do with heat engines, its much wider applicability soon became apparent, and it is now regarded as the most general regulator of natural activity known to science. The second law governs the way in which energy and matter can be exchanged between systems in an ordered way, and effectively forbids us from using the same quantity of energy over and over again for a useful purpose such as to run a machine.

In a nutshell, the second law states that disorder can never spontaneously give rise to order. More precisely, this law legislates nature's trading account in a quantity called entropy, which is, roughly, the degree of disorder in a physical system. When it comes to engines, entropy is related to the amount of useful energy available. In any process, we lose control of some energy; it dissipates away into the environment. When ordered energy becomes disordered in this fashion, the entropy rises. The second law forbids the entropy of a complete system from falling. Even the most efficient machine cannot claw back the heat wasted as friction.

Now it might be supposed that the variety and complexity of natural processes is so great – the forms of energy and matter and the nature of their activity being legion – that at least one instance of a violation of the second law might be found. Not so. Whenever new types of matter or interactions come along, they are always found to comply with the second law.

Take, for example, gravity. This is a topic that seems to have no direct connection at all with thermodynamics. Nevertheless, a curious imaginary experiment devised by Hermann Bondi reveals otherwise. (In what follows I have re-designed the apparatus somewhat.) Figure 29 shows a see-saw (teeter-totter) made from a stiff optical fibre. At each end of the fibre are spheres containing a single appropriately chosen atom and the external surfaces of the rod are light-sealed with silver. Suppose initially the atom in the left-hand sphere is excited. Having more energy than the atom on the right it will weigh more, and so gravity will try to pull the see-saw down on the left and up on the right. The force of this motion can be used to drive a dynamo which powers a machine. Eventually the rod reaches a maximum inclination, which for best efficiency would be vertical, with the excited atom at the bottom (Figure 29(*b*)). The machine now stops.

So far, nothing very remarkable has happened. At this stage we recall that excited atoms are unstable and will eventually decay to an unexcited state by emitting photons. When this happens to the excited atom in the lower sphere, a pulse of light will travel up the fibre. On arrival at the top sphere it will excite the atom contained therein, making it heavier than the atom at the bottom. The rod will therefore be top-heavy and start to swing round again, continuing until the excited atom is once again at the bottom and the unexcited atom is at the top. Further power is extracted during this process. The whole cycle is then repeated, *ad infinitum*.

Although the forces involved here are miniscule, and the power output is hardly a sensation, in principle the device seems capable of generating unlimited quantities of energy for nothing, if you are prepared to wait long enough (or have enough duplicates). It is a modern version of the *perpetuum mobile* so ardently sought by medieval inventors as the perfect answer to the energy crisis. As such the device comes into conflict with the second law of thermodynamics which forbids *perpetuum mobiles*. But where is the flaw?

Careful analysis shows that there is a hidden assumption involved in the operation of the device. This assumption is that no change takes place in the excited atom as it swings down from the higher location to the lower. But this is not correct. We have forgotten one of the effects of gravity. As explained in Chapter 2, gravity slows time. The excitation of an atom is

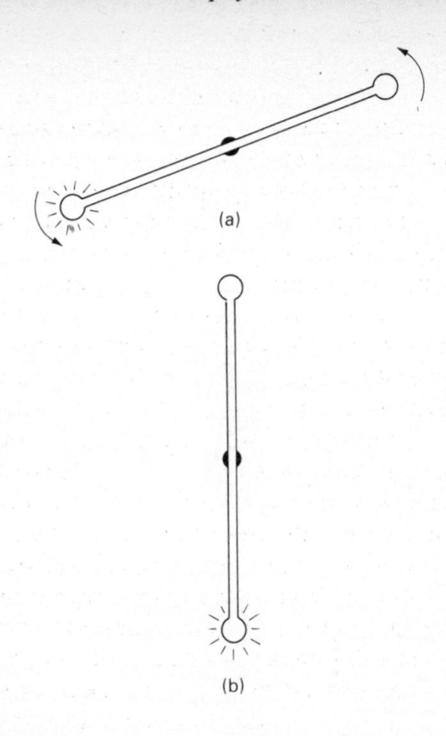

Figure 29. *Perpetuum mobile?* The pivoted rod is an optical fibre with a ball, each of which contains one atom, on each end. (*a*) If the left-hand atom is excited it weighs more, and so the rod begins to tip. In principle, energy can be extracted from its motion. (*b*) Eventually the rod will come to rest vertically with the excited (heavier) atom at the bottom. Thereafter the atom may de-excite and emit a photon, which climbs up the fibre to excite the atom in the top sphere. This leads to an overbalance, which causes the rod to start rotating again, enabling more energy to be extracted. If the effects of gravity on time are overlooked, this device seemingly violates the laws of thermodynamics by providing an unlimited source of free energy.

rather like a vibration; thus, if time is slowed, so is the frequency of vibration. This implies that the energy of excitation is reduced somewhat, and it is precisely this energy loss that is being syphoned off to run the machine. When the photon climbs up the optical fibre, therefore, it will arrive at the top with less energy than before, and will either fail to excite the atom there, or produce only a lower level of excitation. After a few cycles, the excitation energy will be negligible and the device will come to a halt. The second law of thermodynamics triumphs again.

In discussing this intriguing example, Bondi pointed out that the slowing of time by gravity is one of the fundamental principles on which Einstein's general theory of relativity is based. It can be shown to imply the well-known fact, enunciated by Galileo, that all freely falling bodies accelerate equally fast. If we did not know these facts about gravity already, we could deduce them, from a law of thermodynamics.

We have already seen in Chapter 3 how a similar consistency between gravity and quantum mechanics enabled Bohr to rescue Heisenberg's uncertainty principle from an attack by Einstein. What beautiful examples of the consistency of physics!

About fifteen years ago, physicists thought they had finally run up against a physical system that was weird enough to escape the strictures of the second law. This system was the black hole.

The first systematic investigation of the thermodynamic properties of black holes was carried out in about 1970 by Jacob Bekenstein, now of the Ben-Gurion University of the Negev, then a graduate student at Princeton, following a suggestion of Robert Geroch of the University of Chicago. Bekenstein conceived of a 'thought experiment' in which a box full of heat radiation was slowly lowered on a rope towards the surface (known as the horizon) of a black hole, whereupon it was opened and its contents sacrificed to the hole, the box being withdrawn again to a safe distance (*see* Figure 30).

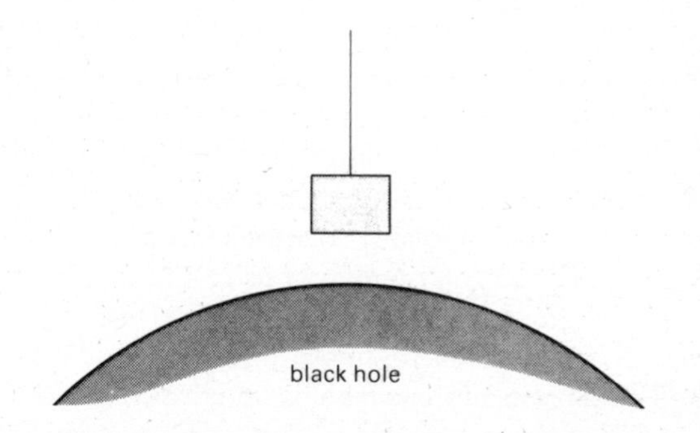

Figure 30. A box full of heat radiation is slowly lowered towards the surface of a black hole, whereupon a trapdoor is opened and the heat energy falls into the hole. There seems to be a paradox concerning the second law of thermodynamics, because the hole swallows up the entropy of the heat but not its energy, the latter being extracted gradually as a result of the work done by the descending box.

The apparently irreversible loss of heat down the black hole reduces the entropy of the surroundings, and so Bekenstein conjectured that the hole must itself carry entropy which rises as a consequence of its eating heat, thus saving the all-important second law. Spotting that any sacrificed energy inevitably causes the black hole to grow larger, he hit upon the idea that the area of the hole's horizon (roughly its surface area) is a measure of its entropy.

These speculations were placed on a firm foundation by Stephen Hawking of the University of Cambridge, who in 1974 announced the spectacular result of a new mathematical analysis. Hawking applied the quantum theory, normally reserved for atoms and molecules, to the subject of black holes, and came up with the first of a long list of surprises. Black holes are not black at all, he found, but clothed in a glow of heat radiation. As befits a consequence of an essentially atomic theory, the Hawking radiation is important only for microscopic holes of nuclear dimensions but, significantly, it endows every hole with a new form of entropy that confirms Bekenstein's original guess: the area of the hole is the relevant quantity. Energy may enter the hole from outside and, by waiting for the Hawking radiation to stream off, it may eventually be returned to the environment again. In all these exchange processes, the total entropy – ordinary plus black hole area – should never decrease.

How well does the new, generalized second law of thermodynamics work? In simple energy exchanges the black hole's entropy undoubtedly soaks up the deficit caused by the irretrievable loss of ordinary entropy down the hole. But a re-run of the box-on-a-string 'experiment' runs into trouble. The trouble arises from the fact that as the box is suspended at ever lower positions close to the horizon, the effective energy of its contents is depressed by the hole's overpowering gravity. The reason can be traced to the work expended by the weight of the contents as it is lowered. So strong is the hole's gravity that as the horizon is approached, the total energy content of the box (including the contribution of its rest mass, computed using $E = mc^2$) dwindles to zero. It follows that if a door in the box is opened and the box contents are tipped into the hole, then the energy delivered will be substantially less than that originally incarcerated in the box.

The significance of this energy deficit is easily appreciated. The size of a black hole is determined by its total energy content: add energy and it grows bigger in proportion. The entropy of the hole also depends on its size, i.e. the area of the horizon. Adding energy therefore boosts the hole's entropy. The problem with the box-on-a-string scenario is that the above-mentioned energy deficit implies less entropy boost for the hole. Bekenstein found that if the box is opened very close to the horizon, the effective heat energy is so depleted that it can't buy enough black hole entropy – enough, that is, to offset the entropy carried down the hole with the heat radiation. The second law is violated and the way lies open for the construction of a *perpetuum mobile*.

In a comprehensive analysis of the whole issue, William Unruh of the University of British Columbia and Robert Wald of the University of Chicago came up with the resolution of the problem. The essence of the Unruh–Wald argument is that a crucial influence on the box – the quantum aspects of the hole – has been left out of the equation. From a distance the hole appears shrouded in heat radiation due to the Hawking effect. Although for a large hole the temperature is negligibly small, the *effective* temperature experienced by a suspended box rises steadily as it is lowered towards the horizon.

The rise in effective temperature can be envisaged heuristically as follows. The escalating timewarp caused by the hole's gravity implies, crudely speaking, that time runs slower and slower as the hole is approached, grinding to a complete halt at the horizon; all relative, of course, to a distant clock. Heat radiation, consisting of waves, contains countless natural clocks – the wave pulsations – which are forced to tick ever more frenetically in the dilated time at greater depths to keep pace with their more elevated counterparts and maintain thermal equilibrium. The higher frequencies imply higher temperatures. Thus, in a gravitational field, thermal equilibrium implies a temperature gradient. As the Hawking radiation has precisely this equilibrium character, one expects it to be hotter nearer the hole.

Armed with this new feature, Unruh and Wald soon discovered that the behaviour of the suspended box is drastically altered. To confine its own heat radiation, the box must have highly reflective walls. But the same property that keeps its own radiation in also serves to keep the Hawking radiation out. Consequently, as the box is lowered, it punches a cavity in the heat shroud that clothes the hole, and the resulting displacement of radiation induces an upthrust on the box in exactly the same way as the displacement of water keeps a boat afloat. Though Archimedes would probably turn in his grave, it is his celebrated principle being invoked here.

The buoyancy force experienced by the box alters the whole energy–entropy competition, because the effective weight of the box is progressively reduced with depth, and therefore the work released by the descending contents will be diminished. It follows that there will not be as large an energy deficit when the box is opened as previously supposed. In fact, when the box probes deep enough, the soaring temperatures will eventually be capable of completely neutralizing the weight of the box contents. No advantage is gained by lowering the box beyond the neutralization point. When the box is opened and the contents sacrificed, the energy delivered to the hole will be a minimum if the box contents are released at the neutralization point. Unruh and Wald demonstrated that these effects are just enough to save the second law of thermodynamics by imposing a definite maximum on the amount of energy deficit acquired by the box contents during the lowering process.

A more spectacular possibility arises if the box is lowered beyond the neutralization point. The rising buoyancy will eventually reach a level where it will be able to support the entire weight of the box; the string may then be cut and the box will float of its own accord in the hole's heat bath!

A still more exciting possibility arises when an empty box is lowered to the neutralization point and opened, for it will instantly become filled with high-temperature radiation from the heat shroud that clothes the hole. This heat energy may then be withdrawn and used. We will have literally mined energy from the hole (Figure 31).

This achievement would appear totally paradoxical without the concept of 'negative quantum energy', for ultimately the energy must originate with the black hole itself, yet nothing, including energy, is supposed to be able to escape from a black hole. However, it can be shown that the energy which appears in the box has been acquired, not by straightforward extraction from the hole, but by injecting negative energy into the hole. This negative energy influx will cause the hole to shrink a little to pay for the heat in the box. Nevertheless, the hole can soon be refuelled by dropping in an equivalent amount of waste mass. We thus have available, in principle, a device capable of converting any unwanted matter into heat energy.

Nobody is suggesting that the Unruh–Wald discovery will solve the world's energy problems, or that it even remotely corresponds to reality. The box-on-a-string scenario is a fantasy, a thought experiment designed to test the validity of the laws of physics. But this does not eliminate its importance. If the basic principles of thermodynamics, quantum theory, and gravity were not compatible, even in an imaginary situation, we should be compelled to abandon at least one of them. The fact that consistency is achieved in such exotic circumstances gives us welcome confidence in the universal validity of these fundamental laws.

The message behind this story is that the black hole brings together

Figure 31. Mining a black hole. An empty box is lowered towards the hole. A trapdoor is opened, enabling the box to fill with the intense heat radiation associated with the hole. The box is then withdrawn, and the heat energy put to use. The energy is paid for by a stream of negative energy emitted by the reflective underside of the box down into the black hole. This reduces the energy (hence mass and size) of the hole. The process therefore effectively mines the energy locked up in the hole's mass.

three rather different branches of physics: there is gravity, which is needed to make the hole in the first place; then there is quantum mechanics, which makes the hole glow and emit heat radiation; and finally there is thermodynamics, which regulates the exchange of energy between the hole and its environment. At first sight it appears that there is a clash. The black hole appears to violate the second law of thermodynamics. In fact, it turns out not to be so, but only when quantum physics is taken into account. The three branches of physics mutually support each other, even for a system as bizarre as a black hole. Furthermore, the consistency of these very different topics is only revealed in a subtle way, after some careful examination of rather unusual effects (such as the floating box phenomenon) that were certainly not obvious at first sight.

The black hole provides a good example, then, of how physics hangs together coherently, sometimes for the most subtle of reasons. Had we not known about quantum mechanics and had only the laws of gravity and black holes at our disposal we should have concluded that something was very wrong. Conceivably we might have been led to invent the Hawking radiation phenomenon, and gone on from there to deduce the rules of quantum physics.

Natural genius

There is a story that Newton made an elaborate clockwork model of the solar system. When someone remarked on how clever he was to construct such an intricate mechanism, Newton replied that the Good Lord must be a lot cleverer still to have constructed the real thing.

Who cannot fail to be struck by the ingenuity of the natural world? Nature is astonishingly clever in the way it gets things to work. The entire superforce saga is a classic example of ingenious mechanism. Consider the business of gauge symmetries and of getting forces out of the need to maintain symmetry under general gauge transformations. A less ingenious Mother Nature would just have put the forces in 'by hand'.

Then contemplate the unification of the forces. How clever and elegant for all the forces needed to make our complex and interesting world to be generated out of just one superforce. Again, nature could have chosen the more brutish method of just giving us four straight forces. As if all this wasn't enough! The fact that the whole gauge-field structure is mathematically precisely what is needed to describe the world in terms of the pure geometry of eleven dimensions – itself a unique structure with unexpected and most specific mathematical properties – seems like a miracle.

Just as miraculous is not only what nature has given but what it has left out. The four forces are just enough to build a world of modest complexity. Without gravity, not only would there be no galaxies, stars, or planets, but the universe could not have even come into being, for the very notion of the expanding universe, and the big bang as the origin of spacetime, is rooted in gravity.

Without electromagnetism there would be no atoms, no chemistry or biology, and no heat or light from the sun. If there were no strong nuclear force then nuclei could not exist, and so again there would be no atoms or molecules, no chemistry or biology, nor would the sun and stars be able to generate heat and light from nuclear energy. Even the weak force plays a crucial role in shaping the universe. If it did not exist, the nuclear reactions in the sun and stars could not proceed, and supernovae would probably not occur, and the vital life-giving heavy elements would therefore be unable to permeate the universe. Life might well be impossible. When we remember that these four very different types of force, each one vital for generating the complex structures that make our universe so active and interesting, all derive from a single, simple superforce, the ingenuity of it all literally boggles the mind.

What is equally remarkable is that although all four forces are necessary for a complex and interesting world, nature has not decided to 'play it safe' and throw in a couple more for good measure. This astonishing economy – just enough to do the job and no more – prompted the British mathematical physicist Euan Squires to ask, 'Do we live in the simplest possible interesting world?' Squires concluded that a universe which permits some sort of chemistry, and hence life, could not be constructed from forces and fields of a simpler nature than those we actually perceive.

Physics is full of such examples of ingenuity and subtlety, and one could fill volumes discussing them. One final illustration will, I hope, suffice to convince the reader that nature is very, very clever indeed. The example also illustrates the themes of unity, order and harmony.

Fundamental to the notion of an orderly world is some degree of permanence. If the whole world changed erratically from instant to instant, chaos would reign. We want to be confident that the parked car stays parked, that the furniture stays put when arranged, that the Earth does not shoot off into interstellar space, and so on. The property of matter 'staying put' is so basic to our experience that we rarely question it. The world would be horrific indeed if bodies flew off of their own accord without any motivating agency.

We can generalize this slightly because we know that a body can only be stationary in one reference frame. More generally, a body will move in a straight line, without acceleration, in the absence of an applied force. This elementary fact is embodied in Newton's laws of motion. To use Newton's own words: 'The description of right [i.e. straight] lines . . . upon which geometry is founded, belongs to mechanics.' The question we want to address here is how the body achieves this miracle, so vital to the orderliness of the world. How does it know what path to follow? How does it work out what is a straight line?

The answer has its origin, believe it or not, in quantum effects and, in particular, in the wavelike nature of quantum particles. This is a topic we have already encountered, in Chapter 4. It had long been known from the study of optics that light rays also travel in straight lines. There is, in fact, a very close correspondence between the motion of a material body and the motion of a light ray, even in more complicated circumstances where forces are present and the paths are curved. In essence, both comply with what might be termed the principle of indolence: they move in such a way as to minimize their total activity. (The level of activity can be defined precisely in a mathematical way that need not trouble us here.) In a sense, the light ray and the material body both take the 'easiest' route available. Remember that a straight line is the shortest distance between two points. Yet light consists of waves, whereas a material body is a discrete particle or collection of particles.

This unity of principle between waves and particles is suggestive of a deep harmony in nature concerning motion. But the way in which nature achieves the straight-line motion of the material body is brilliantly ingenious. At the quantum level a particle does not follow a precise trajectory at all, certainly not a straight line. Instead, its motion is fuzzy and chaotic. How can we build the orderly straight-line motion of a macroscopic body out of the chaotic quantum behaviour of its component atoms? Here, nature seems to turn a sin into a virtue. As explained in Chapter 2, a quantum particle gets from A to B by 'feeling out' all possible paths simultaneously; remember how the single photon somehow goes through both slits in the Young's interference experiment. More generally, we can imagine a particle such as an electron exploring all the various convoluted paths that connect its point of departure with its point of arrival (*see* Figure 32). By a principle of democracy, each path contributes equally to the total wave that represents the electron, and which encodes the probability that it will arrive at a certain destination.

It is at this stage that the vital wave nature of the electron plays a part. As discussed on page 28, when waves are superimposed, 'interference' results. If the waves are in step, they will re-inforce, if they are out of step they will cancel. When a large collection of waves are superimposed in a random way, the effect is wholesale cancellation. This is precisely what happens with all the convoluted paths followed by the electron. The waves associated with those paths essentially annihilate each other by destructive interference. The only paths for which this does not occur are those where all the waves just happen to be in phase with each other, and hence re-inforce rather than cancel. Exact enhancement only occurs along the straight-line path, and to a limited extent in the paths close to it. Hence the particle most probably follows the shortest route available. The degree to which a particle is likely to wander in an indeterminate way from the straight and narrow is determined by its mass. For an electron, the motion is highly erratic and ill-defined, but a heavier particle is less adventurous. In the limit of a large body such as a billiard ball, deviations from a straight-line path are infinitesimal. In this way the well-defined, straight-line trajectory of

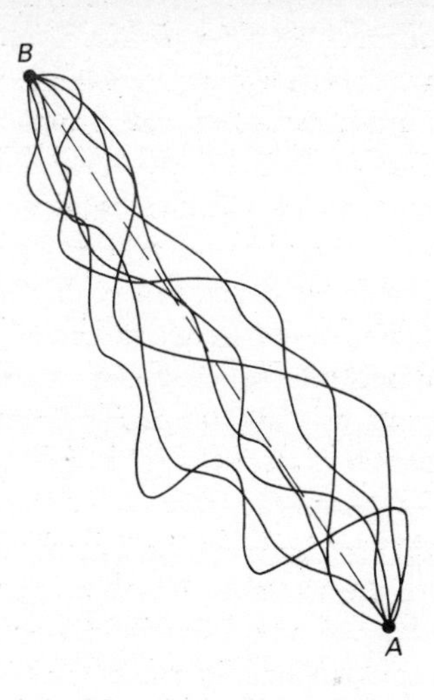

Figure 32. How does a particle 'know' which is the straight path from *A* to *B*? Quantum theory provides the answer. The particle simultaneously 'feels out' all possible paths between *A* and *B*. Because of the wavelike nature of quantum matter, destructive interference causes wave cancellation everywhere except in the region of the straight path (broken line). Hence, according to the probability interpretation of quantum waves, the most probable paths are those concentrated near the straight line. Only at the atomic level do we notice any significant deviation from the straight-line, classical trajectory.

classical mechanics is recovered. Thus, the origin of the orderly behaviour of macroscopic bodies may be found in the quantum physics that ultimately underlies them.

Design in the universe

A common reaction among physicists to remarkable discoveries of the sort discussed above is a mixture of delight at the subtlety and elegance of nature, and of stupefaction: 'I would never have thought of doing it that way.' If nature is so 'clever' it can exploit mechanisms that amaze us with their ingenuity, is that not persuasive evidence for the existence of intelligent design behind the physical universe? If the world's finest minds can unravel only with difficulty the deeper workings of nature, how could it be supposed that those workings are merely a mindless accident, a product of blind chance?

Once again, the crossword puzzle analogy is appropriate here. Uncovering the laws of physics resembles completing a crossword in a number of ways. Nature provides us with 'clues', often cryptic, and the solutions to the puzzles usually involve subtleties. The laws are not manifest in a casual inspection of the world. Rather, they are hidden behind the more obvious activity and can be discovered only by digging beneath the surface. The laws of atomic or nuclear physics would never be apparent without special technology and carefully designed experiments. Nature confronts us with something like the cryptic clues of a crossword. Solving the clues requires considerable ingenuity, practice, and inspiration, for the answers are rarely obvious.

When several 'clues' have been solved, a pattern begins to emerge. As in a crossword, where the words interlock in a consistent and orderly arrangement, so the laws of nature interlock consistently, and we then begin to discern the remarkable orderliness of nature alluded to earlier in the chapter. The world is a conjunction of physical mechanisms. This conjuction does not lead to a haphazard jumble of effects, as might be so easy, but to a carefully organized harmony.

In the case of the crossword, it would never occur to us to suppose that the words just happened to fall into a consistent interlocking pattern by accident, that the subtlety and ingenuity of the clues are merely brute facts of no significance, or the product of our own minds attempting to make sense of meaningless information. Yet one frequently encounters precisely these arguments concerning the miracle of nature, which is overwhelmingly more subtle and ingenious than any crossword. If, then, we do not doubt that the order, consistency, and harmony of a crossword imply that the puzzle is the product of an ingenious, inventive mind, why are such doubts voiced in the case of the universe? Why is the evidence of design so compelling in one case but not in the other?

In the nineteenth century the existence of order and harmony in nature was frequently used by theologians as an argument for a supernatural designer. One of the most articulate proponents of the design argument was William Paley, who employed an analogy between natural mechanisms and a *watch*. Paley invited one to consider coming across a watch unwittingly and, after examining its intricate mechanism of interlocking components, reasonably concluding that it had been designed for a purpose by some intelligent mind. Comparing the watch with the many extraordinarily refined mechanisms in nature, such as the orderly arrangement of planets in the solar system and the complex organization of living creatures, Paley declared that the evidence for intelligent design was still more forcefully apparent than in the case of the watch.

In spite of its superficial appeal, Paley's argument – and many subsequent attempts to deduce the existence of design from the workings of nature – has been savagely attacked by philosophers and scientists. Three of the rebuttals that are still deployed today are as follows: that we impose order on the world to make sense of it; that the reasoning is flawed; and that any order which does exist in nature is the product of blind chance and not of design.

First, we impose order on the world to make sense of it. The point here is that the human mind is most adept at finding patterns amid a tangle of data, a quality which presumably confers evolutionary advantages on us. We are continually presented with complex information which has to be organized in some way by the brain in order for us to function effectively. The famous constellations, where random distributions of stars in the sky were perceived as coherent patterns by our ancestors, is a good example of the mind perceiving order where none exists. There is no Big Dipper, no Virgo, no Scorpio; there are only haphazard points of light.

Nevertheless, the argument is not wholly convincing when applied to science. There are objective ways of determining the existence of order in a physical system. The order of living organisms, for example, is clearly not a figment of our imagination. When it comes to fundamental physics, the laws of nature find expression in mathematical structures which are often known to mathematicians well in advance of their application to the real world. The mathematical description is not simply invented to give a tidy description of nature. Often the fit between the world and a particular mathematical structure comes as a complete surprise. The mathematical order *emerges* as the physical system is analysed.

A good example is provided by the eleven-dimensional description of the forces of nature. The mathematical 'miracle' that the same laws which govern the forces can be expressed in terms of some previously obscure geometrical properties of a multidimensional space must be considered amazing. The order that is being revealed here has not been imposed, but has emerged from lengthy mathematical analysis.

No physicist would seriously believe that his subject matter was in fact a disorderly and meaningless mess, and that the laws of physics represented no real advance of our understanding. It would be ludicrous to suppose that all science is merely an artificial invention of the mind bearing no more relation to reality than the constellation of Pisces bears to real fish.

The second rebuttal is that the reasoning is flawed. It is sometimes objected that the existence of design in the universe is based on the fallacy of *a posteriori* reasoning, or 'thinking backwards'.

Consider, for example, the following passage from the book *Life Beyond Earth* by Gerald Feinberg and Robert Shapiro:

'A geographer with a predestinist viewpoint might eventually be struck by just how fit the Mississippi River was for its valley. It flows in exactly the right direction, with exactly the needed contours and tributaries, to ensure the drainage of the waters of the central United States into the Gulf of Mexico. In doing so, it passes conveniently by every wharf and under every bridge in its path. The geographer might then attempt to replace the Mississippi, hypothetically, with the Amazon River. Superimposing the Amazon onto a map of the United States, he might notice at once that it flows west to east. This would not work, as it would have to flow over mountains. Even when he turned the river in the "right" direction, he would notice many difficulties. New Orleans would be flooded by the large Amazon delta, and an endless number of roads and towns would be submerged. He would conclude that the Amazon was unfit, and the Mississippi eminently fit, for its purpose.

'Let us restrict the situation further. Assume that the geographer had no knowledge of other river systems but had studied the Mississippi extensively. He would notice that any major change in the river's shape would cause damage and dislocations, and conclude that this shape was the only one possible for a functioning geological system. If other rivers existed, they must have the same general shape.'

Similar criticism is contained in a recent article by Ralph Estling in *New Scientist*:

'Underlying the argument for the supernatural or the super-intelligent is the anthropic principle, the realisation that the Universe is so *exactly* the right kind of Universe for man that we must meditate on the thousands of coincidences that are absolutely essential for man, or indeed life, to exist. One slight variation in just one of those thousands of essential coincidences would have altered the physical Universe drastically, possibly totally. Yet, down to the fine structure constants that dictate gravitational, electromagnetic, and strong and weak nuclear forces, and up to basic biological prerequisites, we find the cosmos in general, our Sun in particular, and Earth most particularly, so minutely attuned to *us*, that the conclusion seems inescapable: God or someone else of the same name made it like that, with us in mind. It is, we insist, just too much of a coincidence, just too much of a miracle, to say it is pure, unnecessitated chance.'

These authors quite rightly draw attention to the pitfalls of thinking backwards, but one must not conclude that it is always fallacious. It is easy to find instances where it works splendidly. Indeed, we employ such reasoning all the time in daily life without coming unstuck. Paley was certainly correct in his assumption that a watch is a product of design. The point is that we have to be careful to avoid the indiscriminate use of *a posteriori* reasoning.

How do we know when backwards thinking is likely to lead us astray in relation to order in the world? The key criterion is to distinguish between two quite distinct forms of order. This brings me to the third objection against design, that any order which does exist in nature is the product of blind chance and not of design.

This powerful objection is undoubtedly correct in many cases, and was largely responsible for the abandonment of the design argument by theologians. However, it is often applied indiscriminately, without distinguishing between two very different meanings of the concept 'order'.

One meaning of order is *complex organization*, as exemplified by living organisms. Consider, for instance, the human eye. This delicate and intricate mechanism seems to be exquisitely designed for the purpose of providing us with sight. The arrangement of the lens and retina is perfectly ordered to comply with the principles of physical optics. The many millions of cells that make up the eye and optic nerve are each highly specialized to perform a particular function, and to co-operate with their neighbours in a controlled and orderly way. A random collection of cells, let alone a random collection of atoms, could never achieve the 'miracle' of sight.

Biologists don't deny the incredible level of adaptation displayed by the eye, or any other organ. Nevertheless, they do not need to suppose that the eye has been designed in advance and assembled by supernatural means. The theory of evolution provides a perfectly satisfactory explanation for how the human eye came to exist. The fossil record and comparative anatomy provide a detailed picture of the way in which a complex organ such as the eye can develop in stages over very many generations in response to evolutionary pressures. Random genetic changes – the product of blind chance – generate all manner of possibilities, and those that confer advantages upon an organism are selected by nature in the continual struggle for survival. A species will explore a vast range of possible alterations before hitting, quite by accident, on one that will improve its adaptation to the environment.

Complex organization can therefore arise spontaneously, without the need for any pre-ordained plan or design. Central to the success of such procedures, however, is the existence of an ensemble, by which I mean a large collection of similar systems. In the biological case, the thousands of millions of organisms and millions of generations which have existed throughout the history of the Earth constitute an ensemble. The huge pool of similar genes provided by the multiplicity of organisms enables nature to 'experiment' with all manner of possible alternatives until, by chance, a favourable mutation happens to come along. Selection then isolates this mutation and 'fixes' it into the gene pool. The accumulation of countless such advantageous small changes establishes, in a slow progression, mechanisms as complex as an eye.

In contrast to the concept of order as complex organization, there is the order involved in symmetry and simplicity. This sort of order can be both spatial and temporal. A good example of the former is provided by the crystal lattice. In a crystal the atoms lock together in a regular array forming a simple geometrical pattern with a high degree of symmetry. This underlying atomic pattern is reflected in the symmetric shapes that the crystals tend to display, such as the cubic shapes of salt crystals. It is this atomic symmetry that is ultimately responsible for the regular shapes of snowflakes. A second example of spatial order is the arrangement of the solar system, in which nearly spherical planets revolve in nearly circular orbits around a nearly spherical sun.

In both these examples we can trace the origin of the spatial order to symmetries in the underlying laws of physics which control the systems concerned. Many physical systems have stable states which display a high degree of symmetry and simplicity. Of course, we still have to explain how the systems get into such states in the first place. One reason is that complicated states tend to be unstable. The lowest energy state of the hydrogen atom, for example, is spherically symmetric, whereas most of the excited states are not. Similarly, the equilibrium shape of a gravitating fluid body (without rotation) is a perfect sphere. We have seen how it is a universal law of nature that physical systems seek out their lowest energy states. If a system begins with excess energy (i.e. excited states) all sorts of mechanisms generally exist to rob it of energy. Sooner or later it will settle into the lowest energy state, which is generally the simplest. For this reason, spatial order is a common feature of the world. It is important, however, to remember that it owes its origin to the spatial order already contained in the laws of physics. If, for example, the force of gravity were more complicated, and depended upon the orientation of two bodies as well as their separation, the planets would follow much more erratic orbits.

Let us turn now to temporal order. This is exemplified by the regularity of many natural processes: the ticking of a clock, the vibrations of an atom, the pattern of day and night, summer and winter. Again, these regularities can be traced to the underlying laws of physics, which frequently permit simple *periodic* behaviour. Indeed, periodic motion, or oscillation, is perhaps the most widespread example of order in physics. Wavelike oscillations lie at the heart of all quantum motion; electromagnetic waves carry heat and light across the universe; planets, stars, and galaxies all involve objects moving on periodic orbits through space.

In addition to the orderly motion of material bodies there is a deeper sort of temporal order in the world which is implicit in the very notion of the laws of nature, and which is often taken completely for granted. The fact that there are laws at all implies a certain consistency in the world from one moment to the next. At its most basic level this consistency is simply that the world continues to exist. Moreover, the laws do not change from epoch to epoch (they would not be regarded as laws if they did). The Earth follows an elliptical path around the sun today just as it has done for millions of years.

Neither spatial nor temporal order is merely an incidental feature of the world; both are built into the underlying laws themselves. It is the *laws* which encapsulate the astonishing orderliness of the world, rather then the actual physical structures. These laws are doubly remarkable because they permit *both* the order of spatial and temporal simplicity *and* the order of complex organization. The same set of laws which give rise to the simple shapes of crystals also permit systems as complex and intricate as living organisms. One could certainly envisage a universe where the laws were so constructed that simple patterns of behaviour, such as the regular motions of the planets, were permitted, but in which no exceedingly complex structures such as polymers, let alone DNA could exist. Indeed, it seems truly extraordinary that such simple laws as those exposed by modern physics permit the variety and complexity of the real world. Yet such is indeed the case.

Is there a meaning behind existence?

It is interesting to ask just how improbable it is that the laws of physics permit complex structures to exist. How finely must these laws be 'tuned'?

In a famous article in the journal *Nature* British astrophysicists Bernard Carr and Martin Rees concluded that the world is extraordinarily sensitive to even minute changes in the laws of physics, so that if the particular set of laws we have were to be altered in any way the universe would change beyond recognition.

Carr and Rees found that the existence of complex structures seems to depend very sensitively on the numerical values that nature has assigned to the so-called fundamental constants, the numbers which determine the scale of physical phenomena. Among these constants are the speed of light, the masses of the various subatomic particles, and a number of 'coupling' constants such as the elementary unit of charge, which determine how strongly the various force fields act on matter. The actual numerical values adopted by these quantities determine many of the gross features of the world, such as the sizes of atoms, nuclei, planets, and stars, the density of material in the universe, the lifetime of stars, and even the height of animals.

Most of the complex structures observed in the universe are the products of a competition or balance between competing forces. Stars, for example, while superficially quiescent, are actually a battleground in the interplay between the four forces. Gravity tries to crush the stars. Electromagnetic energy fights against it by providing an internal pressure. The energy involved is released from the nuclear processes legislated by the weak and strong force. In these circumstances, where a tightly interlocking competition occurs, the structure of the system depends delicately on the strengths of the forces, or the numerical values of the fundamental constants.

Astrophysicist Brandon Carter has studied the stellar battleground in detail, and he finds that there is an almost unbelievable delicacy in the balance between gravity and electromagnetism within a star. Calculations show that changes in the strength of either force by only one part in 10^{40} would spell catastrophe for stars like the sun.

Many other important physical structures are highly sensitive to minor alterations in the relative strengths of the forces. For example, a small percentage increase in the strength of the strong force would have caused all the hydrogen nuclei in the universe to have been consumed in the big bang, leaving a cosmos devoid of its most important stellar fuel.

In my book *The Accidental Universe* I have made a comprehensive study of all the apparent 'accidents' and 'coincidences' that seem to be necessary in order that the important complex structures which we observe in the universe should exist. The sheer improbability that these felicitous concurrences could be the result of a series of exceptionally lucky accidents has prompted many scientists to agree with Hoyle's pronouncement that 'the universe is a put-up job'.

The supreme example of complex organization in the universe is life, and so special interest attaches to the question of how dependent is our own existence on the exact form of the laws of physics. Certainly, human beings require highly special conditions for their survival, and almost any change in the laws of physics, including the most minute variations in the numerical values of the fundamental constants, would rule out life as we know it. A more interesting question, however, is whether such minute changes would make *any* form of life impossible. Answering this question is difficult because of the absence of any generally agreed definition of life. If, however, we agree that life requires at least the existence of heavy atoms such as carbon, then quite stringent limits can be placed on some of the fundamental constants. For example, the weak force, which is the driving force behind the supernovae explosions that liberate the heavy elements into interstellar space, could not vary too much in strength from its observed value and still effectively explode stars.

The upshot of these studies seems to be that many of the important physical structures in the universe, including living organisms, depend crucially on the exact form of the laws of physics. Had the universe been created with slightly different laws, not only would we (or anybody else) not be here to see it, but it is doubtful if there would be any complex structures at all.

It is sometimes objected that if the laws of physics were different, that would only mean that the structures would be different, and that while life as we know it might be impossible, some other form of life could well emerge. However, no attempt has been made to demonstrate that complex structures in general are an inevitable, or even probable, product of physical laws, and all the evidence so far indicates that many complex structures depend most delicately on the existing form of these laws. It is tempting to believe, therefore, that a complex universe will emerge only if the laws of physics are very close to what they are.

Should we conclude that the universe is a product of design? The new physics and the new cosmology hold out a tantalizing promise: that we might be able to explain how all the physical structures in the universe have come to exist, automatically, as a result of natural processes. We should then no longer have need for a Creator in the traditional sense. Nevertheless, though science may explain the world, we still have to explain science. The laws which enable the universe to come into being spontaneously seem themselves to be the product of exceedingly ingenious design. If physics is the product of design, the universe must have a purpose, and the evidence of modern physics suggests strongly to me that the purpose includes us.

Superstrings
A Postscript

Such is the pace of modern research that since the hardcover edition of this book went to press a further major development has taken place in the unification programme. It goes by the name of *superstring theory*.

In the conventional approach to modelling the world, all matter is composed of particles, and the search for the fundamental particles forms a central motivation for the study of high energy physics. As we have seen, even the fields that express the forces of nature receive a particle interpretation in terms of 'messengers'. Now this basic assumption has been challenged. The world, it seems, might not be composed of particles at all, but *strings*.

String theory began in the early seventies, with attempts to understand the internal states of hadrons. It turns out that quarks tied by gluons whirling around inside hadrons behave in some respects like spinning strings. While suggestive, this early string theory was not completely successful. In particular it was found that under some circumstances the strings could move faster than light, which is absolute taboo. The topic became a bit of a backwater while most theorists turned elsewhere, but the theory was kept alive, most notably by Michael Green of Queen Mary College, London University and John Schwarz of Caltech.

Then, in the mid 1970s, a significant development took place that eventually transformed the tricky old string theory into something incomparably more powerful and elegant. The concept of supersymmetry was by that time exercising a strong influence on theoretical particle physics, and theorists investigated the consequences of making strings supersymmetric. It became clear that the new 'superstrings' had enormous advantages over the old strings. For a start, the problems about superluminal motion were eliminated. Secondly, the low energy limit of the theory looked rather familiar – it bore a strong resemblance to *supergravity*. It began to appear as if superstrings might contain much more than a theory of hadrons.

What really made the physics community sit up and take notice, however, occurred in 1983, and concerned a remarkable mathematical property of superstrings that appeared almost 'too good to be true'. One of the mathematical diseases that tends to afflict quantum particle physics is something called the anomaly problem. Anomalies is the somewhat innocuous name given to mathematical terms in a quantum theory, which should be zero on account of very basic symmetries already built into the theory before its quantization. In other words, quantizing the theory leads to the gratuitous appearance of terms which have 'no right to be there'. Such terms, moreover, spoil the consistency of the theory, and can lead to such undesirable predictions as the violation of the laws of conservation of energy and electric charge. The astounding feature of the particular version of superstring theory then being investigated by Green and Schwarz is that there is an unexpected conjunction of mathematical terms which just happens to cancel the anomalies exactly! 'Things you'd never imagine would cancel, do,' declared Mike Green. Miraculously, the theory is anomaly-free

The anomaly cancellation miracle was enough to attract the attention of some other very distinguished theorists. But this was just the beginning. The cancellation is found to occur only if the superstrings are constructed with a very particular form of gauge symmetry, known technically as SO(32) or E_8XE_8. Unlike in particle-based theories, where one is free to build in many contending forms of gauge symmetry, a consistent superstring theory almost uniquely pins down the permitted gauge group. And the fact that the two allowed groups contain the familiar groups, such as SU(3), associated with the weak, strong and electromagnetic forces, itself already indicates a semblance of standard low-energy particle physics showing through.

The final factor that boosted superstrings to overnight fame concerned the fact that the theory has to be formulated in *ten* spacetime dimensions. In the past, the higher dimensionality of string theory seemed to make it hopelessly unrealistic, but after several years exposure to Kaluza–Klein theory, the physics community was able to embrace the idea of higher dimensions with equanimity. After all, the unwanted extra dimensions can always be got rid of by 'compactification'.

The key point about ten dimensions, however, is that it enjoys a major mathematical advantage over the eleven dimensions of Kaluza–Klein theory, for it was proved by Ed Witten of Princeton that any theory formulated in an odd number of dimensions suffers a serious shortcoming. The issue has to do with the existence of 'handedness', or chirality, in nature. As we have seen, the weak force introduces a left-right asymmetry into physics, and it is only if one starts out with a theory in an even number of dimensions that one can end up with a four-dimensional chiral universe. This obstacle, which is a very serious one for Kaluza–Klein theory, is evaded in ten-dimensional superstring theory.

The essence of the strings' advantage over particles is their high energy behaviour. At low energies the strings behave much like particles, but as the Planck energy is approached the internal string motions start to become important. This drastically alters the mathematical structure precisely where the conventional theory starts to go wrong by encountering unwanted infinities. By a happy combination of supersymmetry and internal string motion, it seems very likely that these infinities can really be eliminated at long last.

Superstring theories, which grew out of modest attempts to model certain properties of hadrons, have thus assumed the status of a fully-fledged unification programme. They fall into two varieties, those in which the strings are open-ended and those where they form closed loops. Green and Schwarz originally favoured the open-string version, but this form admits only the SO(32) symmetry group. Some theorists find the other group, E_8XE_8 more attractive, partly because it is possible to construct the theory as, in effect, pure gravitation, and to recover the other forces from it, as is done in Kaluza–Klein theory.

The letter E here stands for 'exceptional', and is so-named by mathematicians because the existence of such groups is mathematically rather unobvious. The closed string models make use of a double copy of E_8 (hence the designation E_8XE_8) which leads to a curious possibility. It seems to predict the existence of two distinct worlds, one for each copy of E_8. The particles within each world would enjoy all the usual properties, including the ability to interact via the various forces of nature. The particles in the 'other' world would have their own identical, but separate, version of these forces. Thus, no direct interaction would occur between particles in one world and those in the other, save in one case – gravity. The gravitational effects of the other world's matter would still show through.

This leads to the eerie idea of a 'shadow universe' interpenetrating our own, going largely unnoticed. There could, for example, be shadow matter passing through you at this very moment, its feeble gravity too weak to cause noticeable effects. On the other hand, a shadow planet plunging through the solar system could throw the Earth from its orbit. And a shadow black hole would be indistinguishable from one formed out of 'ordinary' matter. Significantly, cosmologists have long known that there is a great deal of invisible matter in the universe, causing gravitational disruption but otherwise remaining completely inconspicuous. The possibility that some of this invisible matter might be 'shadow matter' has not been overlooked.

Witten has described superstring theory as 'miracle, through and through' and confidently predicted that it will dominate physics for the next fifty years. Whether this euphoria is premature remains to be seen, but meanwhile much work is left to be done. As with Kaluza–Klein, superstrings constitute a 'top-down' theory, which means one starts with a superforce – a sweeping and elegant unification of all the particles and forces of nature at ultra-high energies – but must somehow 'project down' to the messy bread-and-butter physics of the real world. A way has to be found to go from strings in ten dimensions to the low energy behaviour of particles in four dimensions if the theory is ever to make contact with experimental physics. At present, the mathematical subtleties of this step seem formidable. Nevertheless, the attraction of what have become known as Theories of Everything – complete unification programmes such as Kaluza–Klein and superstrings – is proving so compelling that there are many talented theorists eager to have a try.

Further Reading

(* indicates a more advanced treatment)

General reading

CALDER, Nigel. *The Key to the Universe: a report on the new physics* (BBC Publications: London, 1977).
CAPRA, Fritjof. *The Tao of Physics* (Random House: New York/ Wildwood House: London, 1975).
CAPRA, Fritjof. *The Turning Point* (Simon & Schuster: New York, 1982).
DAVIES, Paul. *God and the New Physics* (Dent: London/Simon & Schuster: New York, 1982).
FEYNMAN, Richard. *The Character of Physical Law* (MIT Press: Cambridge, Mass., second edition, 1982).
MARCH, Robert H. *Physics for Poets* (McGraw-Hill: New York, second edition, 1978).
* MEHRA, J. (ed.). *The Physicist's Conception of Nature* (D. Reidel: Hingham, Mass., 1973).
MORRIS, Richard. *Dismantling the Universe* (Simon & Schuster: New York, 1984).
PAGELS, Heinz. *The Cosmic Code* (Michael Joseph: London/Simon & Schuster: New York, 1981).
TREFIL, James S. *Physics as a Liberal Art* (Pergamon Press: Oxford, 1978).

Particle physics

* CHENG, D. C. and O'NEILL, G. K. *Elementary Particle Physics* (Addison-Wesley: Reading, Mass., 1979).
CLOSE, Frank. *The Cosmic Onion* (Heinemann: London, 1983).
DAVIES, P. C. W. *The Forces of Nature* (Cambridge University Press: Cambridge, 1979).
FEINBERG, Gerald. *What is the World Made of?* (Doubleday: New York, 1977).

MULVEY, J. H. (ed.). *The Nature of Matter* (Clarendon Press: Oxford, 1981).
POLKINGHORNE, John. *The Particle Play* (W. H. Freeman: San Francisco, 1980).
TREFIL, James S. *From Atoms to Quarks* (Charles Scribner's Sons: New York, 1980).

Cosmology

ATKINS, P. W. *The Creation* (W. H. Freeman: San Francisco, 1981).
BARROW, John and SILK, Joseph. *The Left Hand of Creation* (Heinemann: London/Basic Books: New York, 1984).
DAVIES, Paul. *The Runaway Universe* (Dent: London/Harper & Row: New York, 1981).
GRIBBIN, John. *Genesis* (Dent: London/Delacorte: New York, 1981).
* HARRISON, E. R. *Cosmology* (Cambridge University Press: Cambridge, 1981).
* SCIAMA, D. W. *Modern Cosmology* (Cambridge University Press: Cambridge, second edition, 1982).
SILK, Joseph. *The Big Bang* (W. H. Freeman: San Francisco, 1980).
WEINBERG, Steven. *The First Three Minutes* (Basic Books: New York/Andre Deutsch: London, 1977).

Index